STEALING GREEN MANGOES

STEALING

GREEN

MANGOES

TWO BROTHERS, TWO FATES,
ONE INDIAN CHILDHOOD

SUNIL DUTTA

An Imprint of HarperCollinsPublishers

STEALING GREEN MANGOES. Copyright © 2019 by Sunil Dutta. All rights reserved. Printed in the United States of America. No part of this book may be used or reproduced in any manner whatsoever without written permission except in the case of brief quotations embodied in critical articles and reviews. For information, address HarperCollins Publishers, 195 Broadway, New York, NY 10007.

HarperCollins books may be purchased for educational, business, or sales promotional use. For information, please email the Special Markets Department at SPsales@harpercollins.com.

FIRST EDITION

Designed by Renata De Oliveira

Library of Congress Cataloging-in-Publication Data has been applied for.

ISBN 978-0-06-279585-4

19 20 21 22 23   LSC   10 9 8 7 6 5 4 3 2 1

The jasmine blossom does not know how far
its fragrance will travel
Why ask the lovers to know the outcome of
their passionate frenzied hearts?

—SA'IB

———

THIS BOOK IS DEDICATED TO MY
BELOVED WIFE, WES. A LONG TIME AGO
OUR PATHS CROSSED ACCIDENTALLY IN
INDIA. WHAT AN INCREDIBLY FABULOUS
JOURNEY IT HAS BEEN!

Renounce the world, renounce hereafter,
Renounce God, Renounce renunciation;
Yearn instead for a life free of desire!
—HAZRAT KAMIL

CONTENTS

STEALING

GREEN

MANGOES

INTRODUCTION

Before anything, there was God,
Had there been nothing, there would have been God.
It was because I lived that I died,
Had I never lived, what would have been?
—GHALIB

I still do not know whether it was divine providence or sheer misfortune to have been born a refugee. I was born a few years after the partition of India and the following chaos in which upwards of one million Hindus, Sikhs, and Muslims killed one another. My parents and their extended clan survived at the expense of being uprooted, chased out of their ancestral lands, and hated for being Hindus in West Punjab, and therefore part of a religious minority. On the other side of the border, in East Punjab, the Muslim minority received the same treatment from the Hindus and the Sikhs.

The peace that followed brought its own madness. Frenzied fanatics and gross profiteers exploiting religious division killed dozens of my family members. The rest of us were scattered across India. My father's family was left close to penniless, my mother's only slightly better off. Some of us never saw one another again; some reunited only accidentally, decades later.

This is the world into which I and my brother, Kaushal—affectionately nicknamed Raju—were born. It was a world of bitter shame. My relatives, desperate for some semblance of their more exalted past, fought viciously with one another over property and position. Their squabbles were the background noise of our childhood. Their hopelessness made them lash out at Raju and me, and we knew abuse as well as poverty.

My brother and I faced exactly the same adversities. Raised in the same small room, we slept side by side and grew up steeped in the same history, fear, deprivations, anger, and resentments of a refugee family. But we made very different decisions thereafter. Raju and I have lived very different lives.

I turned inward as I grew up. I read. I became interested in poetry, music, and metaphysics. I found solace in study. I sought out mysticism and idealism before finding myself (perhaps paradoxically) in a laboratory, working as a research biologist. Then I transformed my life once more: I became a police officer on the streets of Los Angeles. And I wrote.

Raju also underwent transformation. His life took him into Pakistani terror camps, to Canada with Khalistani terrorists, to France, and then to prison. He destroyed many lives, including that of a woman he somehow enticed to marry him while on the run from the police, and a son, who died of cancer while Raju was in prison for murder.

The subject of this book is the possibility of a life full of compassion and meaning in a broken world. The challenges and dilemmas of my family will be the material of this story, but its lesson is not specific to my experiences. What I hope to convey, by pairing my life with my brother's, by juxtaposing the violence I have seen on the streets of Los Angeles with the violence I knew growing up as a refugee (and with the violence sowed by my own brother) is that events and tragedies do not define or shape our lives. I did not feel compelled to replicate the destructiveness and hopelessness of the family I grew up in, but Raju did. In between Raju and me is the world.

In June 2016, I also learned that I had terminal cancer. Suddenly the urgency to assess my and my brother's life took on a new meaning. I knew that metastatic lung cancer had the lowest survival rates of any cancer; less than 5 percent of those with this disease live five years after the diagnosis. Most die within seven months of their diagnosis. But that is statistics, not my life. This book is also an effort to confront my imminent destiny and a spiritual quest to seek meaning in our existence. It is also an examination of the tortuous paths our lives take and our reactions to events in the face of which we may feel powerless.

We may turn our eyes away from the past, but the past never leaves us. I felt it imperative to examine and assess the paths my brother and I took in life—the decisions we made, the courses of action imposed upon us. Everything a human does is connected with his past. One who lives an examined life reflects, makes connections, and looks for the lessons learned—it is a ceaseless, sleepless quest.

1

NEWS FROM FRANCE

November 17, 2004, Southeast Division, LAPD

Sufis say that God made the human body from clay but that the spirit refused to enter this inanimate prison. God then infused the clay structure with music and rhythm. The spirit was ecstatic and joined the body. Thus humans came into existence. As I saw blood slowly dripping from a child's body on the pavement, I wondered whether his spirit had ever heard the divine music in the many prisons of his world—walls encircling other walls in a never-ending concentric circle. Prisons of poverty and racism—crime-ridden neighborhoods where futures looked dim and life was cheap.

It started as a pleasant sunny Southern California afternoon. Sammy and I were on our way to the Southeast Division, the dreaded South-Central Los Angeles—the South-Central notorious for its gangs, drugs, and violence; known around the world for race riots in 1965 and 1992; its streets and alleys made famous by numerous Hollywood movies, including *Training Day, Colors, Boyz n the Hood* . . .

Sammy and I were both Internal Affairs investigators. We were southbound on the Harbor Freeway in our unmarked police car when I heard our radio crackle. The three short beeps signaled an emergency, quickly followed by the operator coming on air. "All Southeast units, shooting in progress, 109th and Figueroa . . ."

"Knuckleheads started shooting a bit early today; it's only two P.M." Sammy was nonchalant as we exited the Harbor Freeway at Century Boulevard. Shootings were the evening rituals of violent neighborhoods, making the evening shift the busiest.

"I haven't been to a shooting call in two years." Because I was working as an Internal Affairs investigator and had been out of the field almost two years, I almost sounded wistful. Patrol is where the adrenaline floods the body; impossible situations, unpredictable people, speed, fun, and terror come together. I had been away too long.

In the Los Angeles Police Department, shooting calls did not cause as much excitement in the southern parts of Los Angeles as they did in the Valley. Police divisions in South Los Angeles had to deal with much more violence and many more homicides than those in the Valley part of Los Angeles. Before my IA stint, I had spent most of my time working in the Valley divisions; Sammy, on the other hand, had worked only South-Central beats.

If you could put complete opposites together in a cop car, we were it. Sammy was an Iranian Jewish immigrant. His family had escaped Khomeini's Iran and found refuge in Los Angeles. His deep love for Persian culture was matched by his equally deep suspicion of anything Arab or connected to Islam. Sammy was a dedicated cop, as loyal to the LAPD as he was to his own family. A political conservative, he also was a gun enthusiast. Sammy spoke fluent Farsi and had worked deep undercover jobs, successfully penetrating Iranian crime syndicates. And yet none of this could shield him from the ethnic jokes that other LAPD cops made at his expense. People automatically presumed that he was a Muslim. An aquiline nose on a professorial face lent Sammy a scholarly aura. He was large and out of shape, and I doubted he could sprint after a suspect to catch him. He compensated for this by his expert marksmanship and demeanor. I had a reputation of being a political liberal—exactly the kind of person Sammy did not like. Where I was too careful and by the book, he was more impetuous and relied on instinct. Although we were at the opposite ends of the political spectrum, we got along well. Perhaps it was nothing more than our shared outsider status in the LAPD or our love for big dogs and nature, or maybe ideologies don't matter much in friendship.

As two Internal Affairs investigators on our way to a police misconduct investigation, we should have ignored the call. I was so overloaded with police misconduct complaints that I hadn't even found the time to prepare for today's case. My lieutenant had saddled me with fifteen police misconduct investigations while other investigators had a caseload of six. I hated being unprepared for an interview, but today I would have to prep minutes before talking to a witness.

The operator came back with an update. "Southeast units, shooting in progress, four victims down, 109th and . . ."

"Eighteen George . . . we will be Code 6 in a minute, requesting additional units . . ." Southeast was LAPD patrol station number 18; gang units were assigned the designation *George*. A Southeast gang unit had broadcast that it was very close to the scene. This could turn very interesting and dangerous if they ran into the shooters.

"Did you hear that? You wanna go?" Sammy asked.

"Yes, but I have that interview in thirty minutes, and I've been chasing this sergeant for weeks. I can't delay my investigation." I was divided. The IA investigators were discouraged from getting involved in street policing. No cop liked to be around IA investigators anyway. But the call was less than a minute old; the killers were probably still nearby. Also, if there were four gunshot victims, there might be four separate crime scenes, overwhelming the officers rushing there. They could definitely use two extra bodies at the crime scene.

"Let's go," I told Sammy.

"We'll be there in two minutes. It's only twenty-five blocks." He was not joking! Sammy floored the accelerator and the car barreled down the street toward Figueroa.

"Slow down—you'll get us killed. We don't even have lights and siren!" I screamed.

"Come on, do you have to talk like that?" Sammy feigned hurt feelings. He was nearly as proud of his driving as of his Distinguished Expert shooting medal.

I strained my ears, awaiting a description of the suspects. An LAPD helicopter flew over us heading south. Suddenly I felt a rush of adrenaline. Patrol is thrilling and addictive. Maybe we'd see some action.

"There it is, there it is!" Sammy yelled. When excited, his voice turned into a high-pitched squeal.

An angry crowd had gathered near the northeast intersection of the street, screaming and waving their arms; some were throwing gang signs while others looked despairingly and were crying. Three officers were desperately trying to keep the crowd from rushing into the crime scene, pushing people back and away from the crime scene tape. Yellow crime scene tape blocked off the entire street. The Southeast Division covered a small area of Los Angeles with a disproportionately high murder rate. The officers in this division were experts at quickly sealing off the homicide scenes to preserve the evidence. I jumped out of the car and lifted the tape. Sammy parked near the other police cars. Behind us, the all-black crowd was cursing the three young white officers who were still struggling to keep the crowd away from the crime scene.

My eye caught sight of two officers on the sidewalk near a wrought-iron gate. The gate led to a run-down single-story apartment complex. Like most buildings in the area, it was a nondescript gray structure with small, boxy apartments. Unsure about what had happened and who was in charge of the incident, I strode toward the nearest officers.

The young and athletic white officer stood next to a skinny, wobbling, dreadlocked black teenager. The teen had a blank look. His arms hung limp by his side. The officer was holding the teen up by his upper arms, supporting him in the same manner as parents use to help their babies learn to walk. Ignoring Sammy completely, the officer started briefing me. I realized Sammy was in a suit and I was wearing the LAPD uniform. We were the first supervisors to show up on the scene, so I was expected to take charge and ensure that things were running smoothly; but I felt uneasy about treading on Southeast Division's territory.

"Hi, Sarge, the one over there is dead. This one with me was shot in the leg. He's okay. There are two more KMA [dead] in the complex." The dreadlocked boy stared through me with contempt. His tattoos and clothing marked him as a gangbanger, the street term for a gang member. Blood had soaked his thin left calf and his white sock had turned red. He was trying to mask the pain with bravado. The officer was helping him stay up as they waited for an ambulance.

"Any information on suspects?"

"No, sir. It's the usual. He says he didn't see anything: he doesn't know what happened or who shot him." The usual means just that: in the Southeast Division, every witness always claims not to have seen or heard anything. In this area, even the victims don't tell the police who shot at them or killed their friends. Gang members value self-reliance and dispense their own swift justice outside the criminal justice system—no need for witnesses, evidence, or trials.

Nobody talks to the police in South L.A. Especially not in broad daylight, in front of dozens of people. Three young men had been killed, and no one had seen anything. The gang member who survived the shooting while his friends lay dead in a pool of blood refused to even acknowledge the existence of officers trying to help him. It wasn't just him. Even people who didn't hate the police—and there were quite a few who lived in these violent neighborhoods—couldn't be seen talking to us. They had to live and get along every day in a territory ruled by vicious criminals.

Suddenly I recoiled in horror. As the officer spoke to me, I had failed to notice that I stood just a few feet away from the lifeless body of a large black man sprawled facedown on the sidewalk. He was six feet tall and weighed more than 200 pounds. My eyes focused on his head. He had been shot at such close range that the muzzle flash had singed the hair around the spot where the bullet had entered. A score had been settled. This was an execution, not some cowardly spray-and-pray drive-by shooting where gang members driving through a rival gang's territory unleashed a barrage of gunfire, hoping to find the right target. The shooter had stepped up to the victim and finished him off with a contact shot.

"Have we locked down the scene at the rear?"

"Yes, sir. George units are in the back with other bodies."

I looked up and saw my academy classmate Brett entering the complex. He looked relaxed, smiling as he spoke into his radio. He waved at me and disappeared into the building. Brett was tall, brutally strong, and built like a WWF wrestler. Brett and two other academy mates of mine, Nagle and Hatfield, had become gang officers in the Southeast. These days, the entire black leadership and the civil rights establishment in Los Angeles were after Hatfield's scalp. Live national television news cameras captured him beating a car thief with his flashlight after a car chase, an event that was compared to the Rodney King beating. Eventually Hatfield ended up getting fired from the LAPD because of the incident.

I hadn't seen Brett in seven years and wondered if working in South Los Angeles had made him callous. Why was he smiling while walking toward homicide victims? I had often heard officers speak of crimes involving gang members killing their rivals as "victimless." I wondered what it would take for me to cross the line that separated human beings with compassion from insensitive hardened cops. I hadn't quite reached that point yet, but understood quite well what constant exposure to remorseless sociopaths could do to police officers.

The wails of the sirens on the fire department engines and ambulances grew louder as I looked down at the large dark body pressed against the asphalt. As some fleeting thoughts about the indignity of death and the transient nature of the physical body as custodian of the spirit wafted through my head, I saw the left hand of the corpse suddenly quiver, pulling me back from the cerebral realm to the material.

"He's not dead, he's not dead! Get an ambulance over here!" I shouted at Sammy.

"The RA [rescue ambulance] is pulling in right now," Sammy responded. He had been quietly surveying the crime scene behind me.

I stepped closer. The victim had feebly lifted his left hand to his head; his fingers were hooked like claws and he was trying to scratch the bullet wound. The movement was slow and delicate. Strength had left his body, along with so much blood. His left cheek was flush to the ground and his lips were quivering. He was trying to say something. I brought my face near his and tried to listen amid all the noise and confusion. I couldn't understand what he was trying to say.

I knew not to take anything for granted in police work. Nonetheless, the victim should have been dead. The wound seemed to indicate that the bullet had entered the base of the skull and traveled upward to the brain. The dime-sized entry wound encircled by singed hair was proof that someone had jammed the barrel next to his head before squeezing the trigger.

"He's still alive! Get the RA!" I shouted again. In my peripheral view, I saw paramedics in blue uniforms rushing toward me.

I wanted to shout in his ears, "Who shot you? Who killed you? You know you aren't going to live—tell me who did it. Give me a name, just give me a name. Come on, who killed you?" Maybe I could get a dying declaration. Maybe at this point he would divulge his killer's identity. His clawlike hand continued to scratch his head feebly.

"Can you hear me?"

Very faint garbled sound escaped from his lips, which had by now stopped quivering. I was pushed away by the paramedics. Within seconds, they had cut off his clothes to check for additional bullet entry wounds and turned him on his back. With cold efficiency, they quickly slid the naked body onto the gurney.

Standing above him, I had envisioned some large and hardened gang member. I was wrong. He was a large, pudgy-faced child, no more than fourteen. His eyes were almost completely closed, a sliver of white visible from under droopy lids; he kept mumbling, his faint voice seemingly emerging from afar. Was he calling for his mother? Telling me his deepest secrets? Crying in pain? I stepped closer and saw his bloody right arm move slowly up; he wanted to scratch the bullet wound. I felt an urge to take his hand and help him touch his head.

The paramedics lifted the gurney and rushed toward the ambulance. Suddenly his right arm dropped, the feeble movement of his chest ceased, and his lips stopped moving. He was dead.

What unforgivable impropriety could this teenager have committed to have his life snuffed like this? What lies in the heart of someone who jams a gun in the back of another person's head and pulls the trigger? Was any honor restored or control established over this street corner by this killing? What on earth caused this appalling behavior? Poverty? Upbringing? Lack of opportunity? Bad parenting? Or are some people born with a mission to cause misery?

"Sarge, you have to move out of the crime scene." The clipped voice of a Southeast investigator pulled me out of my thoughts. By now there were dozens of officers and paramedics swarming around, and Sammy and I were no longer needed. As I got ready to leave, I felt a shock of pain shoot through my shoulder, one of the muscle spasms I'd been having for a couple of weeks. I turned to leave while grabbing the back of my neck. We walked in silence toward our car, leaving three dead teenagers and a chaotic scene behind us.

———

THAT NIGHT I SLEPT FITFULLY, my dreams filled with the images of a child's bloody hand falling lifelessly from the gurney interspersed with scenes in which I was running through dark alleys. I chased furiously after the killers, and as I gained on them, I discovered that I was completely alone—no partner, no radio, no backup. Suddenly my gun slipped out of my hand and landed in a pool of blood.

Late in my life, when I should have been contemplating my academic legacy and training graduate students for a scientific career, I decided instead to pick up a gun and strap it on my belt. It was an unlikely decision. Growing up in a destitute refugee family in India, I had never thought that I would become a police officer. Our lives seemed so hopeless that even to dream of a good life was a stretch. Getting enough to eat used to be an achievement when I was small. With some pluck and a heavy dose of luck, I had somehow escaped the toxic environment I was raised in. My friends and family wondered why an idealistic scientist, a lover of music and metaphysical poetry, a social activist, and an ACLU member one day decided to become a cop. Even *I* wondered sometimes. However, somehow it did make sense to me. For a nonviolent pacifist who admired Gandhi, picking up arms was an extension of idealism. I had chosen an incredibly violent path, fully aware of its ugliness, frights, and heartbreaks, yet it was the logical thing to do. Armed with the tools of violence, one could protect and serve the helpless, hopefully without using force or violence. Even Gandhi approved of violence in defense of the defenseless. I had fully immersed myself in a culture with which I had disagreed before; I was a true outsider in an institution whose philosophy and ideology was contradictory to mine. Yet at the age of thirty-four, I had picked up a Glock pistol with seventeen bullets in its magazine and one live round in the chamber, knowing that I was carrying a tool of death and destruction at my side. Every day I was in the worst neighborhoods of Los Angeles, working with a notorious group of men and women clad in blue uniforms. And when the day was done, the horrors of the streets kept returning in my dreams, the bloody nightmares kept disturbing my rest and churning my innards, yet I couldn't let go of this new life.

THE NIGHT I SAW THE teenager die in front of me, my fitful sleep was mercifully interrupted by the harsh ringing of the telephone. My phone rarely rang; something must have happened to get a call at two A.M. Reaching over my wife, I irritably grabbed the phone, but not without a moment's hesitation. I do not like bad news. I always react stoically to tragedies, making others erroneously believe that I am insensitive, but despite my stoicism, I hate *hearing* bad news.

"Have you heard anything from France?" It was my mother calling from India. The anxiety in her voice was unmistakable.

She was obviously wondering about my brother, Raju, who lived in France. My heart started beating faster. Raju was my only sibling. We had grown up in utter poverty in a small room, sharing a wooden cot strung with itchy jute ropes as our bed in Jaipur, India. What had happened to him?

"Are you there? Have you heard anything from Raju?"

My mother's voice pulled me out of the trance. "No, what happened?"

Half awake, I was running through all the possibilities. My elder brother was either seriously ill or dead. I had not wanted my parents to know that he had been struggling with cancer for years, and so I had kept this secret despite his serious condition, his years-long treatment, a very slow recovery, and a possible relapse.

"Did Christine call you?" my mother asked. Christine was Raju's French wife. She would never call me. She couldn't speak English; neither could my nephew, William. In any case, we hardly spoke. During the past sixteen years, Raju and I had had very limited contact.

"No, she did not call me."

"Have you spoken to Raju? When did you speak to him?"

"I don't know when I spoke to him last. I think it was two or three months ago. What happened?"

So he wasn't dead after all. He must be in trouble. Just the thought of Raju's getting in trouble infuriated me. How could he be so selfish? He had committed a series of crimes twenty years ago in India, making our lives difficult. After all the havoc and pain he had caused in the lives of so many, he should be grateful and penitent. Despite all his criminal activity, he had somehow avoided prison, he had found a loving woman who completely supported him, and he had even survived cancer. He did not deserve such good fortune. Why couldn't he stay out of trouble?

"I called Raju today because he hadn't phoned for many weeks," my mother said. "Christine picked up the phone. She was crying."

"What did she tell you?"

"Christine said Raju had killed two people."

"What?" Suddenly the room felt icy. "What do you mean? Did he kill two people in a traffic accident? What exactly happened?" I was in shock, as horrified as I had been watching the Twin Towers of the original World Trade Center collapse early one September morning on the police academy television.

"I don't understand her. She barely speaks English. All she says is that Raju killed two old people, and she keeps crying."

"Okay, I'll call her and phone you back. Maybe it was a traffic accident. Whatever it was, you can't do anything about it, so don't worry."

I looked at my wife, who was staring at me, her eyes wide open.

"I guess you heard that. Can you call Christine?"

Not only was I not a telephone person, I was extremely uncomfortable with the idea of speaking to Christine. I would rather listen to someone else talk with her on the speakerphone. During the last sixteen years, I had exchanged words with her only two or three times. I did not fully understand why I could not bring myself to be closer to Raju and his family. I had never met my nephew, who was almost sixteen at this point. Though Wes didn't speak French, she dialed Christine's number.

"Hello, this is Wes, your sister-in-law. How are you?"

"Kaushal killed two people. He is in jail."

"What happened? Was it an accident?"

"I don't know. I don't understand."

"Who did he kill? How? Did he drive a car into someone? Did he shoot someone?"

"I don't know. He killed two old people."

Christine's vocabulary was limited to common pleasantries, and neither Wes nor I understood any French. It was a stalemate.

"How are you feeling? How is William?"

"I am very lonely, scared. William is quiet."

"Ask her if she needs money," I said in a low voice to Wes. Raju and Christine had both been stuck in minimum-wage jobs for the last sixteen years. They could barely scrape by, and now with Raju in prison, I was wondering how Christine would manage.

"Do you need money?"

"Yes."

"Can you write us a letter in French? We will have it translated."

"Yes."

Maybe Christine understood, maybe not. We still didn't know what had happened. As I dialed my parents in India, I had no answers about whether we were dealing with a homicide or a traffic accident. Raju never drank alcohol, but he had used strong opiates for his numerous cancer-related complications and back pain.

My mother instantly picked up the phone. "What happened?"

"I don't know what happened. I have asked her to write me a letter. I am also going to call the French police to find out."

"What is going to happen? Are we going to be in trouble? Are the police going to come here?"

This was not wild paranoia on my mother's part. For years my parents had to put up with the heavy-handedness of the Indian police because of Raju. He had robbed his own adoptive father of a large sum of money and become a fugitive. While fleeing from the police, he had continued his crime spree of committing fraud, stealing, forging passports, impersonating others, racking up more criminal charges. Family members of people suspected of crimes are fair game for the Indian police; they are threatened, beaten, extorted, humiliated, and sometimes jailed without any charges. In extreme cases, as I had known to happen in Punjab, a family member could be executed to teach the fugitive a lesson.

"Don't bother yourself. Whatever has happened has happened in France. The French police are not the Indian police. They have no reason to contact you. Don't worry at all. If someone contacts you, ask them to speak to me. Don't say anything to them; don't make any statement. In any case, we don't know what has happened."

"Your father thinks that there is a conspiracy against your brother."

I was furious but stopped myself from bursting out in anger. "He is not that important. He is a nobody. Why would anyone spend *any* time to get Raju in trouble? Tell Father to open his eyes and stop trying to cover up for Raju." I felt cold anger toward my father welling up inside me.

I tried to say a few encouraging words to my mother and hung up.

"You already knew what had happened. That's why your shoulder has been in so much pain," Wes said, giving me a knowing look.

"Come on, don't give me this new-age bullshit," I snapped.

"Okay, don't believe me, but your body got the message weeks ago."

I ignored her comment. "I don't know if it was an accident or he actually killed someone. He is a con man, but killing is something else. Maybe because of all the drugs he is on, he has lost his mind. Let me see what I can find out."

A cocktail of painkillers, antidepressants, and cancer medication can alter a patient's perception. In police work, we sometimes see horrendous violence perpetrated by people whose reality is altered by ingesting or failing to ingest various legal and illegal drugs. The imbalance of chemicals within or the chemicals imbibed by humans keep the police busy.

I was on the phone again, this time to France. I spent several hours talking to the jail officials and the police in Lyon and Bourg, getting no information but learning a lesson or two about French police bureaucracy. Raju lived in Bourg, where the officials stonewalled me. The police officers did speak passable English, but they would not even tell me if Raju was in prison for murder or if he had been arrested or charged with *any* crime. They could share information only with "close family members," and since I was calling from the United States, despite being his brother, I did not qualify.

Obviously they did not believe there was an urgent need for me to find out what had happened.

I was used to being treated with respect by representatives of other law enforcement agencies. A simple mention of the fact that I worked for the LAPD impressed cops almost anywhere. In India, I was let inside airplanes without the usual pat-down searches. That day I couldn't bring myself to use my LAPD connection. It would be humiliating to say, "Hello, I am a sergeant with the Los Angeles Police Department. I want to find out if my brother was arrested for killing two elderly people." I didn't want to pull strings to get such horrible information.

I kept trying to reason with the French officers. "What if he needs to be bailed out? His wife has no money. He has been on heavy medication. What if he needs help? If you make a call to an American police station, we will tell you if someone is in jail and what crime he has been charged with. What is this secrecy about?"

"Sorry, you have to call the Justice Ministry and get permission."

I slammed the phone down. Call the ministry just to find out *if* a family member had been arrested?

I had been on the phone for hours. I smiled wistfully. A favorite theme of the Bollywood movies of my childhood was the life stories of two brothers who have been separated early in life. One of the brothers grows up and becomes a criminal, the other a police officer. What a preposterous storyline. My life had become a third-rate Bollywood movie! How did that happen?

I sat quietly for hours, trying to make sense of the two grave tragedies I had faced within twelve hours. That night, when I had received the call from my mother, I did not know that the multiple homicides in South-Central I saw earlier during the day had been preceded by Raju's murdering an elderly defenseless couple and an invalid woman. Three young black males were killed yesterday for no apparent reason other than pathological arrogance, a skewed sense of "honor," and reliance on violence to demonstrate power and "superiority." At least that is the impression the gang members wanted to convey while explaining away their lethal behavior. Other times, they played the victim, blaming others for their behavior. Was their bravado and reliance on violence simply a mask for a deep sense of shame and an inferiority complex? Or did their ignorance and disregard for their history and culture, combined with hedonistic materialistic behavior, make them lash out at their own community instead of dealing with challenges of life and existence with compassion, dignity, and forgiveness? What could explain Raju's actions?

—————

I RECEIVED A LETTER DATED January 15, 2005, from Christine and had it translated.

She had met with Raju's lawyer and told him she couldn't pay him. She asked me not to pay any lawyer who might contact me on behalf of Raju as Raju's case was hopeless. Raju had killed very rich and much loved people in a small town. Christine suspected that there might be a third homicide victim. She did not think that Raju could make it, even with all the money or the best of lawyers. Christine was physically and mentally broken down and terrified about the future . . .

I stared at Christine's letter in disbelief. My entire world was turned upside down. There was no way to deny this horrifying tragedy; the truth came in the form of a clipping from a French newspaper. The article, accompanied by big bold headlines and a color photograph of one of the victims, included graphic details of the double homicide of a retired couple, Marcel and Marcelle Rapallini, eighty-three and seventy-five, who had been beaten and then shot in the chest and neck. The killer had confessed to the murders. Later, it was discovered that he had also murdered a fifty-nine-year-old invalid, Marie-Claude Gaudet. She was killed with one gunshot in the back of her head. The killer, Kaushal Dutta, knew his victims.

The crisis is deeply personal. In Indian society, the responsibility for a crime lies not with the perpetrator, but with their entire family. Raju killed three defenseless elderly people, friends and acquaintances. That would imply that *I* treasonously killed three defenseless old people in cold blood. The humiliation is mine. It is my shame, my utter failure, *my* responsibility. I might have cut my connections with Raju a long time ago, but Indian society does not allow us to sever the bonds so easily. The entire family must be dishonored; we must all pay a price for his misdeeds. And how can I reconcile my own job with my brother's wrongdoings? How does a police officer deal with a brother who is a serial murderer? I needed answers, but where would I go for advice and support? The family shame must remain concealed. Still trying to comprehend what had happened, I wrote to Raju. I heard back from him. In March 2005, he wrote from the Bourg jail, confessing to the murders. Blaming his doctors for not understanding his condition, harassment by his employer, and being humiliated by people, he claimed that he had lost control and shot three people. There was no death penalty in France, and Raju expected to be sentenced to twenty-two years. His letter was matter-of-fact, did not seem repentant, and did not express any deep regrets.

I read and reread the confessional letter, searching in vain for a rational explanation, for some insight, for sincere remorse. What had sent Raju past the edge? Was it his upbringing? Was it our blood-soaked family history? When and why did he soak up the revenge-based violent mentality of our clan? Did the blood he spilled connect to the rivers of blood my family members had to wade through during the partition of India, before we were even born? Did Raju have a flashback to the days of Khalistani terror-ism, when fanatics killed thousands and thousands in the name of religion and blew up jumbo jets in the air? But the two of us had started out on a path together, were raised by the same parents, grew up in the same house, and were close in age. What had hap-pened?

We are shaped by our biology, history, family, and circumstances. But we also mold our own lives; we create our own destinies. I had to study and assess our past in order to understand whether Raju had shaped his own course or destiny had shaped Raju.

2

RAJA PARK

Heat rose from the asphalt in transparent wispy waves. Even the wind seemed defeated by the summer afternoon. The door to our grandmother and aunts' room was closed. Heavy shades shielded the windows from sunlight and darkened the room.

In the afternoons in Raja Park, human activity came to a halt. People napped inside small, stifling rooms in which ceiling fans exchanged the hot air that had settled on the floor for the sizzling air that had risen up to the ceiling. The world became still—even the ubiquitous flea-ridden mangy mongrels hid themselves in burrows dug in the sand. That's when my brother and I were ready for action. When the rest of the world rested—*that's* when we sought out adventure.

Our summer vacation was spent playing marbles under the blazing sun and running around on asphalt and blistering sand to find things to amuse us. Children born to poor refugees couldn't find much to do in the famous "Pink City," as Jaipur is known. It was the beginning of the 1970s. I was eight years old: a scruffy, skinny boy, who ran around the neighborhood in his cheap gray cotton shorts and flip-flops with straps that kept breaking off. TV hadn't made its way to Jaipur yet. Everyone and everything, including our favorite (and sole) entertainment source, the radio, was asleep during the late afternoon. The last music program ended at three P.M., at which point our spiral of boredom began.

But not that day. That day we had a real mission.

Less than a quarter of a mile from where we lived was a small private school, the oddly named Happy School. Happy School was a small single-story building with fewer than ten classrooms. The school compound had a four-foot-high stone wall around it. The grounds were covered with thorny desert trees, reddish-green amaranth, spiky shrubs, poisonous milkweed, and sharp-bladed desert grasses that only the wandering goats with Teflon-coated tongues and titanium-hard innards could stomach.

And then there were the mango trees. These lured us back to the schoolyard even during summer break. When the plants had angled their leaves away from the merciless sun and the human world was taking refuge from the heat, Raju and I were out casing the Happy School and its mangoes. First we scaled the fence and sneaked inside the compound. We climbed over the thick limbs of mango trees, which offered a little bit of shade to the parched soil. The tree branches were bowed with the weight of hard green fruit. The mango fruit had a loving relationship to the heat—intense heat brought intense sweetness.

But we plundered the trees whether the fruit was ripe or not. If we could pierce the skin with our teeth, we ate it. So we filled our pockets with green mangoes. Beneath us, the earth was littered with unripe fruit that we planned to carry in our shirts.

The heat, our sense of adventure, and our joy over a bountiful fruit harvest had made us oblivious to the risks of theft. We had grown complacent about one threat in particular—the dreaded Lallu, who guarded the school compound. Lallu was perhaps only forty years old, but with his disheveled, mottled, dirty gray dhoti, his drab vest, and his perpetually unshaven face, he looked like a petulant old man. He was mean and vicious—meaner even than our grandmother, who threatened to break our ankles at the slightest infractions. Children climbing over the limbs of fruit trees sent Lallu into a frenzy. We had heard rumors that he threw rocks at children who tried to steal fruit from the Happy School.

And somehow, even from his afternoon nap, Lallu had become aware of our presence. He would have caught us, but his fury had sent us advance warning signals from a hundred yards away. As we were gathering the fruit from the ground, we could first hear the earth trembling under his feet—and then a verbal assault unmatched in my lifetime.

"You motherfucking, sisterfucking bastard thieves. Wait till I catch you, you daughterfucking dogs, you fatherfucking cunt insects. I am going to stick this whole tree up your ass before I beat you to death!" And so on.

He came charging at us like a hyperactive dog. The sight of him kicked off my survival instincts and I began to run like a wildebeest in a nature documentary. I didn't know where Raju was, but obviously he was in flight, too. I could hear Lallu running after us, screaming at the top of his voice.

"Wait, you motherfucking shit-eating ass-sucking shit beetles. Let me get my hands on your ass. I am going to shove dry cow dung up your asshole."

We leaped back over the fence and left Lallu panting and screaming in fury, promising to get us the next time. "You fucking thieves, today you steal mangoes, tomorrow you will rape your sisters and murder your mother. Sisterfuckers, you will be *chors* and *dakus* [thieves and robbers], you will bring shame to your parents, you will spend your lives in jail . . ."

Lallu's rants grew faint as we pulled away from him. We pounded over the hot sidewalk and crashed through bushes. Only when we were a safe distance away did we stop to catch our breath under the shade of a large banyan tree.

We were both panting from exertion; my shirt was thoroughly soaked with sweat and stuck to my chest. Pulling the shirt away from the hot skin, I started blowing at my chest to cool down.

"Shit, what happened to our mangoes?" Raju said.

I patted my pockets; they were empty. The mangoes must have fallen out during our mad dash from Lallu. We didn't have anything to show for our adventure.

"Do you want to walk back and find the mangoes?" Raju asked.

Although my appetite was strong, my survival instincts were stronger.

"No, Lallu is probably waiting for us. He'd *kill* us."

We slowly trudged back toward home. At least we'd escaped Lallu—who knew what he would have done? Not wanting to wake up our grandmother and aunts, we gingerly stepped inside our room.

"Where were you? What were you doing outside at this ungodly hour?"

Our grandmother, whom we addressed as *Bhabhiji*, stood with her hands on her hips, giving us a hard stare. Looking down, I mumbled, "Nothing. We were playing marbles outside."

There was a flash, and suddenly I was screaming in pain. The gnarled old witch had grabbed hold of my right ear and was trying to separate it from my head. My forearms instantly went up to block the slaps raining down on my face. Bhabhiji might have looked older than ancient Egypt, but she moved quickly and was highly flexible; she adapted to my defense by changing the focus of her attack to my back and behind.

"I see—lying, are we? Your rotten behavior has destroyed our honor! You shameless rascals, not even ten years old and already on the wrong path. You will end up in the jail; you will become criminals; your mother and father won't be able to show their faces to anyone. This is what your parents get for giving birth to you— shame and early death! Just wait until your father is back, he will take care of you."

A beating was easier to take than Grandma's shaming. In response to spanking, I could defend myself. I could contort my body and find ways to minimize the pain. But there was no defense against her verbal barbs. I was totally defeated, my spirits crushed. I glanced sideways to look at Raju. As usual, he had avoided the beating by offering me first to the hangman. I was the fall guy while he stood safely behind my back. He looked nonchalant.

Apparently while we had strolled back home, Lallu the snitch had jumped on his bicycle and had already showed up at our house to complain. We had been caught red-handed and then ratted on. I was terrified and racked with anxiety. What was going to happen? We were guaranteed a beating when Daddy returned home, but that wasn't a concern. I was more terrified of Bhabhiji's words about us becoming thieves.

There was nothing worse than being a thief. In our clan, bringing the slightest shame to the family was worse than death, and being a criminal was far worse than bringing shame to the family. I would have preferred to be dead than to be called a thief. I hadn't thought that stealing mangoes would have qualified as thievery. Was Lallu calling the police? Would they take me to the jail and beat me with sticks? Was I going to be a thief and spend the rest of my life in jail? Should I pray and ask various gods to intervene on my behalf?

There, in a small, hot room darkened by heavy drapes and made only slightly bearable by the constant spinning of creaky fan blades, I promised myself never to steal mangoes again and began to pray to the monkey god Hanuman to defend me from the oncoming trouble.

"Why can't you be like Kapoor's children? Or Chopra's? They study all the time, they come first in their class, they never create trouble, they are so quiet one would never know they are around, they respect their parents . . ." Father would regret that we were not like other normal, well-behaved children, children that belonged to his friends. Tortured by inner turmoil and worried about the inevitable impending doom in the evening, I called my brother, "Hey, Raju, what's Daddy going to do?"

He was sound asleep.

IN 1959 MY FATHER HAD arrived in Jaipur, the famous capital city of Rajasthan, to accept a lowly clerical job with the Indian government in the Accountant General's Office. Far from Punjab, famous for its five magnificent rivers and wheat-laden farms, and thirteen years after my parents were made destitute refugees, my father had begun a new life in the desert. He made a home in an emerging and sparsely built housing development for Hindu refugees, incongruously called Raja (King's) Park. A fading indigo seal, stamped on a small yellowing folded piece of paper from the Faizabad refugee camp, which identified my father as a refugee, fetched him a four-hundred-square-yard plot of land for 200 rupees.

Raja Park was nothing much more than an open area dotted by prickly desert shrubs. Thick and gritty sand was its most prominent feature. Its yellow particles could find their way into the most improbable nooks of the house. The golden powder of the desert was everywhere—inside people's underwear, in their nostrils, inside their earwax. Not much else existed in Raja Park at the time except the deep ocean of coarse golden silica. The government could afford to be benevolent and essentially give away this land, because it was a wasteland. Who in his right mind would live in the desert? But to a refugee, an opportunity to call a small piece of land his own was a godsend.

Raja Park was a few miles southeast of Jaipur—a majestic city of palaces built on the hills and studded with ancient forts. Jaipur was known as the "Pink City" because its ancient forts and palaces and bazaars within the old walled city were all painted in a soothing shade of pinkish-red. Its rulers claimed direct descent from the sun god and called themselves Suryavanshi: born of the sun.

We lived outside this famous city as underprivileged new arrivals to the land of privileged Rajputs. Our food, clothing, language, history, traditions, and culture were different from those of the natives of Rajasthan. Even our favorite gods were different. When I was five years old, our small house was still expanding to accommodate our joint family. The street we lived on had three houses and the neighborhood was growing very slowly. With no pocket money to spend, few children to play with, and minimal distractions, Raju and I would often wander across deep sand dunes to trek to the hills nearby. The world's oldest mountain range, the Aravalli Mountains, was within walking distance from our house. They were drab and gray, their granite rocks studded with scrawny, spiky shrubs with stems of a purplish brown-green. An arid forest with amazingly resilient acacia trees, Calotropis, Chenopodium, castor, and a million forms of cacti separated the hills from our home. Despite growing in parched earth, the cacti produced elegant blooms with intensely bright colors and delicate fragrances in summer. Their magnificent blossoms transformed the desolate forest with vibrant and beautiful displays of emerald, white, red, orange, blue, yellow, and purple.

That's if we looked in that direction. In the other direction were the slums.

Less than two hundred yards from our home where Street No. 6 ended, there was a densely populated *jhuggi* colony. The flimsy shacks were made up of clay walls plastered with fresh cow dung and patched with desert reeds. Early morning every day, the slum women went on an expedition to find fresh cow dung. They filled shallow pans with mounds of cow excrement and then molded the refuse into patties, which they later used for fuel. Part of the cow clay was hand-smeared on their huts. My father said that cow dung had mysterious healing powers and kept the homes of slum dwellers clean and free from insects and disease. Even the waste coming from the rear end of the holy cow was considered venerable treasure.

The shacks were so tightly packed together that people could barely walk between them. Scavengers, rickshaw pullers, sweepers, housecleaners, and menial workers populated the colony. They were illiterate and poor—even poorer than us refugees. Their huts were tiny and had no electricity or running water. Hundreds of people, crammed in a small lot fit for only a few dozen, shared the sole water tap provided by the government. There were no toilets, forcing them to walk to the forest before sunrise to do their daily duties. Little slum children, running around naked, were a bit less discreet. The slum was a constant reminder that there were people less fortunate than we were. People we could look down upon and abuse. People we expected to lower their gaze when we passed by them.

I was probably the exception in the neighborhood in that I didn't bully the slum dwellers. This had nothing to do with my compassion for the poor. I was simply afraid of everyone, including the skinny men and women covered with rags, who addressed everyone as *mai baap* (mother, father) or *sarkar* (lord). Not only did I lack confidence, I was easily embarrassed, frightened, and fooled.

Raju, on the other hand, was on a growth spurt. He exuded self-assurance bordering on the cockiness we saw in Bollywood film heroes. Neighbors were already whispering that there was something wrong with him because he looked older than his age. At thirteen, he was already as tall as our father and his facial hair had started sprouting. But he wasn't any kind of menace at this point. He was outgoing and made friends quickly. While strolling the Raja Park streets in the evening, he would lock eyes with strangers, grin, and greet them loudly, while I could only cringe in embarrassment. He refused to be intimidated, he spoke his mind, and he couldn't care less when people disagreed with him.

I looked up to my brother and wished that I could be like him—strong, self-assured, uninhibited, unafraid, and calculating. I was completely the opposite. I avoided confrontation and arguments of all kinds. Only every once in a while did I have crazy moments when I would lose my timidity and be transformed into an audacious fool. People ganging up against the helpless—usually the critters or any such unfortunate soul—infuriated me. As a result, I was frequently on the side of the wretch being pelted with rocks, and ended up getting struck myself. Decades later, I still remember my first embarrassing defense.

I was playing in the large sandy lot on Street No. 7, plucking the white buds of akra and watching the plant's thick cream trickle slowly down its shoots. I noticed a slender streak of white smoke slowly drifting above a small crowd, dissipating near the crown of the large gulmohar tree near our home. People were huddled together in front of my playmate Baby's house. All I could see were heads: turbaned heads; the sari- and chunni-covered heads of women; heads covered with small towels to fight off the sun; and on the periphery, even those of children, who could sense there was a big event going on.

I sprinted toward the pack. My cheap rubber sandals couldn't keep up the pace, and one of the straps pulled out of the thin sole. I tripped, rolled over the ground, and ended up flat on my face.

I lifted my head, praying no one had seen me tumbling on the garbage heap. As soon as I did, all of those thoughts disappeared. My mind went blank as my eyes came across its own.

The unblinking eyes stared back at me. The gaze was powerful and unsettling. I remember wishing to turn my head, but I didn't or couldn't.

They were burning a snake alive. Small yellow-orange flames leapt from his gasoline-soaked body. I could hear the crackling sounds of its charred skin splitting. I could smell its burning flesh.

A tiny pair of eyes held me captive.

Mr. Singh, Baby's father, held a long bamboo pole on which the bright orange, red, and black body of a large snake hung. Despite having been doused with kerosene and lit on fire, he had lifted his head up to look at me.

Of all the onlookers, why did he pick me for this honor?

The snake was neither struggling nor writhing to get off the stick. He seemed curiously detached from his destiny, almost stoic. Decades later, I would see this feeling of detachment in the eyes of inner-city kids in Los Angeles who had killed other kids in their own neighborhoods. I would see a similar detachment from life in the eyes of drug-fogged homeless men wandering skid row. I would see a similar disconnect from their destiny in the eyes of cold-blooded white-power gang members being sentenced to life imprisonment.

"Stop that, stop that. What are you doing? Why are you killing him? Stop that right now, let him go!" I heard myself screaming.

"You crazy fool, have you lost your mind? Get back, you don-key, or he will bite you. Do you want to die, you damn fool?" Spit rained from Singh's lips as he bellowed at me.

Baby grabbed my hand and pulled me away. "Are you crazy? What has gone into you?"

"Why are you killing him? Let him go, that poor snake! Why are you burning him?" My agitation and the words escaping my mouth were unplanned. Had I been thinking, I never would have made a case for anything to a mob of snake-burners.

Baby pulled me farther away and gave me a disbelieving look. "What has gone into your head? What did you eat this morning? Standing so close to the snake! What if he jumped and bit you? Don't you know snakes can fly in the air and bite you even if they are dead?"

"How can they bite if they are dead?"

"You know nothing. Why do I have to tell you everything? You are just a *buddhu* [ignoramus]. Not only a buddhu, you are also a crazy snake lover," Baby taunted. She never bought into the traditional subservient role for Indian women. Maybe because she was only eight years old.

The crackling sound of the fire had now been joined by the snake's scales splitting up. A strange smell wafted through the air. Singh moved close to a small sandpit in front of his house, which was used for preparing slurry of lime and cement for constructing homes. He lowered the stick into the sandpit. I moved closer. Someone had piled up dry leaves and sticks in the pit. The snake's body slid down the smooth bamboo stick and fell right in the middle of his improvised pyre. Large flames leaped up, engulfing the snake completely as he disappeared in the inferno. Singh must have soaked the mound with kerosene.

I didn't understand my reaction. Why did I feel empathy for the snake? Usually I was so afraid of snakes that the simple mention of a snake would cause me to jump and run. And there was no scarcity of these slithering animals where we lived. I felt terrible, both for the snake and myself. The snake was gone, free of his misery, but I was left an open target for taunts by the neighborhood kids. As I sheepishly averted my eyes from Baby, I caught sight of my brother in the mob. I wondered what he was going to say.

"That was so stupid to kill the snake like this. They should have smashed his head like this." He stomped on the ground and squashed the imaginary snake's head under the ball of his foot. "Or at least we could have cut his head off to kill him instantly," he said in exasperation.

He hadn't seen my transgression.

Our early childhood gave no indication of our future direction in life. If anything, Raju appeared confident and carefree, whereas I lacked poise and self-assurance. We were unruly and played hard, and we got in trouble together. But don't normal children behave this way everywhere? Perhaps our family history was the cause for Raju's waywardness. As I observe the ongoing human tragedies of immense magnitude in Syria, Afghanistan, Kashmir, Yemen, Myanmar, and elsewhere, I realize that just a generation ago, my own family suffered through experiences no different from those undergone by the current victims of war and terrorism. It is difficult to even imagine what horrors my family and clan faced in 1947, when the British colonialists departed from India, dividing it into India for the Hindus and Pakistan for the Muslims. We were told these stories many times, and in listening to them, we lived them. Were we shaped by them?

3

THE EXILE

I face a sea of blood with large roaring waves
I am content with it; I know worse could happen
— GHALIB

The siren began softly and gradually increased in pitch and intensity. The blast of sound seemed to last forever. This was no fire or air raid—just the daily reminder to men that it was time to start heading to work. Men of Raja Park took their bicycles off the stand and began pedaling to their offices. Women, after a busy morning, could finally breathe in relief. It was time to visit the neighbors, sit on cots strung with itchy jute rope, enjoy steaming glasses of tea (no one used cups; it was sacrilegious not to serve real chai in a small glass), and loudly engage in their favorite activity: gossip.

Women would congregate and walk to the individual's house whose turn it was to provide tea and snacks. If I noticed my grandmother or other relatives in the clique, I followed the group, pedaling on my rickety tricycle. It was hard to keep up with the bunch of moving saris and Punjabi suits. The rubber footrest covering the pedals on my tricycle had crumbled and disappeared a long time ago. As I tried to pedal faster, my feet would slip on the thin metal rod serving as both the pedal and the footrest. I did not give up just because my feet kept slipping and getting scratched by the sharp metal pedal. My lust for snacks kept me going.

No one paid attention to a six-year-old boy; I was privy to conversations about lazy daughters-in-law, cruel mothers-in-law, thankless children, and the price of vegetables. Even though I was not interested in the gossip, I still remember one common refrain among the ladies: they often talked about "back over there." The realm of back over there, subject to exaggeration at times, was a kingdom of both geography and heart. It referenced the village or town or city in Pakistan where my parents and many neighbors lived before the Partition of 1947. It was where they had left behind fortunes, treasures, large landholdings, cattle, farms, and—most precious of all—their homes. The homes over there were large and magnificent compared to any house in Raja Park or anywhere else, and their sense of loss was deep and enduring. Even though the stories of the past were meaningless to me as a child, my view changed as I grew older. I could see that pieces of the hearts and minds of my family had been left *back over there*. Maybe the bitterness, resentment, and rage exhibited by my dysfunctional family were rooted in the partition and the migration that was forced upon them. Maybe the violence-soaked history of my family had permeated the psyche of Raju. But what happens when victims themselves turn into victimizers? Is there any difference between the violence of the oppressor and the victim?

———

ON JUNE 3, 1947, the departing British colonial rulers had announced the division of the Indian subcontinent into a Muslim Pakistan and a Hindu India, effective August 15, 1947. A boundary commission was given less than six weeks to divide a large landmass with a population of nearly 400 million people. The man responsible for creating the borders, Cyril Radcliffe, had never been to India, had no knowledge of the territories he was about to divide, and did not have any cartographic experience or experience in dividing territories. He arrived in India on July 8 and reportedly estimated that it would take years to divide such a vast territory and create the boundaries. Radcliff finished his work on August 12, two days before Britain transferred power to Pakistan and three days before the transfer of power to India. The ill-begotten boundaries, produced in unconscionable haste, were shoddy and controversial. Worse, the borders were kept a secret until August 17, two days *after* the independence from the British! Two new nation-states had come into existence without knowing their boundaries. Suddenly millions of people discovered that they were on the wrong side of the border! Violence spiraled out of control and turned genocidal.

The partition resulted in one of the most brutal and bloody forced migrations in history, as Sikhs and Hindus were evicted from newly created Pakistan and Muslims were expelled from India. Hindus and Sikhs killed Muslims, and Muslims killed Hindus and Sikhs. The ensuing violence resulted in the massacre of between 1 million and 2 million Hindus, Muslims, and Sikhs; the expulsion of approximately 12 million people from their homes; the abduction and rape of over 75,000 women; and the creation of one of the largest waves of refugees in history. By the time atrocities ended, 15 million had lost their homes. The 15 million included my entire clan.

By a twist of fate, my clan's village, Kanjrur-Duttan, happened to fall seven miles west of the new borders. The day Pakistan came into existence, my father's family had become Pakistanis, Hindus living in a nation created as a homeland for Muslims. However, the Duttas of Kanjrur were neither concerned nor fearful. After all, they were the "rulers" of the village. Indians had seen the rulers change before; the people did not change or move with the new rule. As my father told me, the feeling was that the English used to be their rulers, and if his village was to become a part of Pakistan after independence, the new rulers would be the Muslims. He did not expect violence or conflict between the Hindus and the Muslims. This thought lasted for less than a week: on August 22, 1947, all the family members in town on my father's side were forced to flee. Muslim mobs were killing and pillaging in the neighboring village of Mallah; Kanjrur was next in line. The Duttas fled under the cover of night to escape the bloodbath. The landlords became landless refugees.

The partition of India was an abominable event with far-reaching consequences. The Nazis had to employ the power and authority of an industrialized nation-state, within the context of a large-scale war, to kill over ten million Jews, Gypsies, homosexuals, and others during the course of a few years. The provincial Indians and Pakistanis were more effective killers, massacring over a million people within a matter of four months, without armies, tanks, machine guns, or gas chambers. Scores of my uncles and granduncles were butchered in the ensuing religious violence in which Muslim mobs attacked minority Hindus in newly created Pakistan and Hindu mobs attacked minority Muslims in India. My grandaunts were raped and abducted by the Muslim marauders. Even my elderly granduncle Pooran Chand was killed by a mob, his head brutally separated from his frail frame

Despite hearing about these incidents over and over, I chose to tune them out and remain oblivious of my family history. The stories were macabre; they were the past; why hold on to such devastating memories? As the Persian poet Hafiz once asked, why build a shrine to the past and worship it, wailing and lamenting? Only when I began to search for the reasons behind Raju's behavior did I realize the significance of our violent past. The story of the exile of my father's side of the family is stunning in its tragedy and savagery. I knew in general that my parents' families had suffered great hardships when India was divided. I even knew about the murder and mayhem they had faced. Yet confronting the details was unsettling. Two of the stories left deep imprints in my heart. The story of one of my uncles, Pal, exemplifies the horrors of the times and how the atrocities of the period shaped my parents' and clan's attitude toward others, especially the Muslims. The story of another set of my uncles, Beeru and Buddhu, reveals how some victims lost their humanity during the partition and engaged in monstrous violence against "enemies" who were helpless victims themselves. Perhaps all that they experienced made my family members callous and predisposed toward violence and conflict. In fact, I now notice how my family elders were not that much different from the gangs of Los Angeles: they glorified revenge; they saw violence as a solution to their problems; their skewed sense of honor made them intolerant and belligerent, including toward their own blood relatives.

Throughout our childhood, this matrix of violence, through which my family had passed in 1947, was a constant reminder of who we were and what we had become, though I had somehow managed to create a firewall between our history of violence and my psyche. Perhaps Raju could not separate himself from the intense savagery our clan had faced just thirteen years before he was born. Perhaps he became one of the victims whose rage could be quenched only by the blood of the "enemy." If that was true, what stopped me from following in Raju's footsteps?

————•————

ON AUGUST 15, 1947, the very day of the creation of Pakistan, my granduncle Pal,* the younger brother of my paternal grandmother,

---

* The partition stories in this chapter were re-created with the help of my father and other family elders who suffered through these incidents. Pal, Beeru, and Buddhu are my father's uncles and cousins and my father received these details directly from them.

had managed to safely reach the Lahore train station in Punjab, Pakistan. He breathed a sigh of relief; finally an end to the night of terror was in sight. Pal ran toward the platform where the train to Delhi waited. He would be soon on his way to India. For miles, he'd had to dodge Muslim mobs burning homes and butchering Hindus and Sikhs openly in the streets and alleys of Lahore. The streets near Bhati Gate, a Hindu area, were littered with corpses of his Hindu neighbors and friends. He had survived—so far.

Just the day prior, on August 14, Muslim mobs had cut off water and electricity to the Hindu localities and were torching houses, flushing terrified Hindus out into the searing August heat. The trapped victims faced a cruel choice—embrace death by fire or get sliced by swords and daggers. Lahore, the famous cosmopolitan city of British India, was burning. The Muslim police—all the Hindu officers had been transferred to India—watched with indifference, unless they, too, joined in the carnage. The Muslim police in Lahore served Muslims only.

Pal had tried his escape on August 14 but failed. A tall, slender man, Pal loved to dress in the trendy karakul caps, popularized by Pakistan's founder Jinnah, and in a long silken *salwar kameez*. The traditional kameez, a loose-fitting knee-length tunic with long sleeves, and the salwar trousers, loose-fitting with narrow hems above the ankles, were the traditional attire of Muslims. Dressed in a Jinnah cap and salwar kameez, Pal looked like a Muslim. Ironically, the Jinnah cap and salwar kameez had saved his life. When he had fled from his home the day prior, Muslim bands roving the streets, hunting for Hindu victims, ignored him because he looked like a Muslim, not a Hindu.

Pal, terrified that someone might discover he was a Hindu, tried to blend in with an armed Muslim mob waving green Pakistani flags and hunting Hindu prey. This group had come upon the neighborhood of Mohalla Jalotian, where another large mob processed slowly along. The crowd was loud and jubilant. It was a victory procession in which the triumphant warriors were celebrating the ultimate humiliation they had bestowed upon their enemies: they were parading abducted Hindu women who had been stripped naked. The women walked with heads slumped down, dragging their feet lifelessly. Pal felt an almost uncontrollable desire to grab a sword and start beheading everyone around him. But his desire for self-preservation won out. In a letter he wrote to his family in the last week of August 1947, Pal estimated that about 150 stark-naked women were being paraded by the mob.

Although he had planned to leave his house and take a train to India, Pal was too terrified of the naked dance of violence around him. Instead of pushing his luck, he slipped back into his house. It was too risky to be outside, even while blending in with a Muslim mob. Someone might have figured out that he was a Hindu.

Instead of a safe haven, his home felt like a prison. He continued to be tormented by the fear that his Muslim friends, with whom he had shared joys and sorrows for years, would become his executioners. Unable to sleep, Pal twisted and turned on his bed the entire night.

Early the next day, Pal rushed to the fortress-like Lahore Station, the final major railway depot in Pakistan on the route to India. Hindus and Sikhs saw it as their last hope for escape. Horrified by the apocalyptic scene he saw unfold in front of his eyes on the train platform, he collapsed on the grimy cement floor. There the train bound for Amritsar, India, jam-packed with Hindu and Sikh refugees, was under attack. He watched from the platform as the Pakistani police rained bullets on Hindu and Sikh refugees trapped inside the train.* Outside the train, Hindus and Sikhs were

---

* In his book The Punjab Bloodied, Partitioned, and Cleansed, Ishtiaq Ahmed (Ahmed, 2011) documents an account of a shooting incident at the Lahore Railway Station. Saleem Tahir, describing the soldiers of the Baloch regiment, who were upset at having seen the murder and rape of Muslims in East Punjab by Sikhs, mentioned: "This shooting was being carried out by the soldiers of the Baloch regiment . . .

getting hacked to pieces. Screams of the dying were rising up to the sky. The air was thick with the noxious smell of blood. The entire platform was blanketed with a thick pool of red liquid. The killers stopped to scrounge through the blood-soaked bundles of corpses, pulling rings off stiff fingers, ripping earrings off earlobes.

Pal had to use all his might to get up. His legs quivering, he dragged himself to a wooden bench. Terror-stricken, he could hardly breathe.

Pal heard people shout at him as they ran past, "Why don't you join us and kill the infidels?" Some were armed with daggers; others waved unsheathed swords.

"I will never see my family." Pal was shaking in fear. "How long until someone finds out that I am a Hindu?"

"Pal, how are you?" Pal was startled by the question. Rafi, his Muslim friend, stood smirking in front of him. Pal was chilled. He did not know if Rafi was being sarcastic or serious.

In resignation, Pal stared at the ground quietly. What kind of question was that? How was he feeling watching a bloodbath of fellow Hindus, when he could also be at the receiving end if the killers knew who he was? Was he supposed to feel good that he was still alive only because the killers mistook him for a Muslim? He did not know how to respond. Should I beg for my life? What is the point? Is Rafi going to help me? Is he still my friend or has everything changed now, with people turning from friends, neighbors, and coworkers into Hindus and Muslims?

The screams of the tormentors and the cries of the wounded had faded in the background, masked by Rafi's laughter. The derisive chuckling continued to ring in Pal's ears as he watched Rafi

All its officers and soldiers were Muslims. They neither showed mercy to the Hindu railway staff nor any other Hindu or Sikh, military or civilian." *The date of the massacre was believed to be August 14, 1947. It is difficult to establish whether Pal, the brother of my paternal grandmother, saw the massacre on August 14 or a separate one on August 15. He told his story to my father in Ghaziabad in 1948.*

shuffle away. Pal panicked. Did Rafi's departure signify the end of his hopes? Pal's friend, disregarding Pal's plight, had simply walked away without offering solace or protection.

Pal did not know whether to sprint out of the station or collapse in surrender on the blood-covered platform. As his eyes moved up from his feet, he observed five grim-looking Muslims, weapons unsheathed, walking toward him. Looking determined, they surrounded him. There was no possibility of escape now. Pal stared mutely at a sword, its steel darkened with human blood.

"Brother, what is the matter? You look so terrified." Their soft, genuine looks of concern formed a sharp contrast with the metal blades they held. Pal's salwar kameez and Jinnah cap had saved him yet again; the killers had mistaken him for a Muslim.

"My wife and children are in Amritsar. I was going there to bring them back to Pakistan. While you are doing all this here, what will happen to my children in India? Hindus will kill them there like you are killing Hindus here." Pal's survival instincts had kicked in; he had come up with a convincing story without even thinking.

The men appeared startled and were dumbstruck. They stood quietly for a moment until one of them muttered, "Don't you worry; Allah will take care of everything" and they moved away.

In silence, Pal stared at the dun-colored coaches; the metal tubes had become a bloody grave for hundreds of helpless humans. Grim-faced Bhangis had started to hose down the platform. The Bhangis were untouchable Hindus, a caste that for centuries had dealt with human excrement and corpses. Hindu Gods had created Bhangis to perform lowly work for the higher-caste Hindus. However, they were indispensable even for Muslims, for who else would pick up human waste matter, clean latrines, sweep the streets, and pick up the dead animals? The Bhangis were the only Hindus not being slaughtered by Muslims; their utility was indisputable.

The water was mixing with blood, forcing it toward the tracks.

Pal sat in a daze, his face ashen, his shaking hands clutching his belly. He felt faint. The sickening smell of blood had permeated the air.

Dead bodies were being loaded on trolleys, the piles getting higher as the Bhangis moved across the platform. The luggage carts had become wagons for cadavers. A large mound of corpses near the platform was doused with kerosene and set on fire. The killers had respected the final rights of the Hindus by allowing them to be cremated.

As the flames danced over the pyre, Pal forced himself up. He felt no fear anymore. The unending terror had given way to numbness. He slowly stepped out of the station. Death had played a cruel game with him today. He knew he was not going to die.

While Pal was attempting his escape and witnessing massacres of Hindus and Sikhs in Lahore, Pakistan, others from my family, who had managed to escape to India, were witnessing something diametrically opposite. Just a few days after Pal's extraordinary experience, two more of my uncles, then teenagers, were holed up in a dharmshala—a public lodging place for poor travelers—in Dera Baba Nanak, Punjab, India, about fifty miles east of Lahore and eight miles east from our ancestral village of Kanjrur. Most of my clan had abandoned Kanjrur in the middle of the night and walked to Dera Baba Nanak.

———————

THE AIR IN THE ROOM was heavy and stifling. Everything was still. Beeru, my uncle, who was thirteen years old at the time, looked at the rough walls of the bare dharmshala room; they had not been painted for years. The room lacked furniture, windows, and cupboards. The room was more barren than the desert south of his village. The faint yellow glow of a single candle placed in a corner barely pierced the gloom in the air.

Beeru, his younger brother Buddhu, and their cousins huddled together on a dirty cotton sheet covering a dirtier floor. Beeru, normally a bubbly and talkative child, stared quietly at the creases in the sheet created by the shifting of children on a hard floor. The stark eeriness of the room was enhanced by pale streaks of light peering through the numerous narrow slits in the old and decrepit wooden door. Beeru slowly stood up and walked toward the closed door. His small delicate fingers pushed gingerly on the handle. The door was locked from the outside. Uncle Ram Sahay had ushered all the children in the dharmshala to this small room and locked them inside.

The dharmshala was stuffed with his family and other refugees who had escaped from Kanjrur. The day had been hectic. People had been speaking in hushed tones and moving about busily in the courtyard since morning. Everyone had been on edge. His uncles had been discussing "righting the wrongs." Beeru did not understand what they meant. His grandmother sat in a huddle with his aunts and other village women, speaking softly. As usual, no one paid any attention to the children. Now all the children had been unceremoniously locked in the room without explanation.

Beeru slowly made his way back to his space on the sheet, aware of a roomful of eyes watching his every movement. He bent down and placed his right hand on the sheet, sliding next to Buddhu. Suddenly thunderous battle cries rose outside the house. The sounds were distant but furious and loud. Fearful, some of Beeru's cousins started to cry while others sat motionless, staring at the ground. In the distance, loud shouts for help mingled with battle cries, creating an incongruent clamor.

"What is happening?" Buddhu had gone pale. "Are the Muslims going to kill us?"

Beeru put his hand on Buddhu's arm to comfort him. He felt Buddhu trembling under his hand. "Don't you worry. Nothing can happen to us. Muslims are afraid of us; they know we are brave fighters." Beeru remembered how his cousin Hari bragged about the long line of warriors in their clan. He spoke of Dutta warriors to comfort his brother.

Suddenly Beeru heard firecrackers mixed with the screams. The sharp cracks made his heart pound. Even on a Diwali night he had never heard so many loud firecrackers. Unbeknownst to him, in the distance, the days-old Indian military had joined the massacre, their guns efficiently finishing off the helpless refugees.

Beeru's stomach was churning; a sour taste began to rise in his throat. The screams and battle cries were intolerable. He fell to the ground and started sobbing. The steady reports of bullets, interspersed with blood-curdling battle cries, continued late into the night. Beeru lost track of time. The solitary candle in the corner slowly melted into nothingness; even light no longer shone through the cracks in the door. The gentle tap of raindrops on the tin roof gradually turned into a roaring downpour. Thunder and cracks of lightning drowned out every other sound. Unaware of the dance of death not too far from his room, Beeru fell asleep on the hard floor.

THE FIRST PROPHET OF SIKHISM, Baba Guru Nanak, had spent the last days of his life at Dera Baba Nanak. The great sage, born in 1469, had woven a new faith with threads borrowed from Hinduism and Islam, preaching love, tolerance, and conciliation. He mixed with Muslims and Hindus and visited the holy sites of both religions. He was so revered by Muslims and Hindus alike that the followers of both religions claimed Guru Nanak to be one of them.

This holy place of the prophet, situated near the Ravi River, had transformed from a sleepy village to a camp of destitute refugees fleeing bloodshed in West and East Punjab. Between August 14 and August 29, 1947, thousands and thousands of Hindu and

Sikh refugees poured in, increasing the town's population many-fold in less than two weeks. Scores of residents of Kanjrur were among the refugees. People had brought tales of horrendous atrocities from the newly born Pakistan and were full of hatred. They wanted revenge on the Muslims. In the town where Guru Nanak had preached love and tolerance, intolerance and religious hatred burned in scarred hearts.

Thousands of tired and weary Muslim refugees were also camped near the railway station. They had made the perilous trek from eastern Punjab, now a part of India, abandoning their ancestral homes and farms. No one had told these wretches that they would be chased out of their homes like dogs if India was partitioned. The slogan of "Islam in danger" had helped the Muslim League secure Pakistan as a homeland for their culture. While Jinnah and the Muslim League had achieved their Pakistan, the poor defenseless Punjabi Muslims, fleeing Hindu-majority East Punjab, were paying for it with their lives. The British colonialists had stepped down; with freedom, all hell had broken loose in Punjab. In East Punjab, now part of India, Muslims were the minority among Hindus and the Sikhs. West Punjab, in Pakistan now, had a Muslim majority and Hindus and Sikhs in minority. In the east and in the west, religious violence had spiraled out of control. It was the same story everywhere: the majority community was looting, burning, raping, and killing; the minority community was dying or fleeing, leaving all its possessions behind. Law and order had broken down in the cities, with provisional governments paralyzed and ill equipped to deal with anarchy. Massacres were taking place in Delhi, Lahore, Amritsar, and other towns near the newly created border. In the countryside, the situation was even worse.

Muslim refugees in Dera Baba Nanak were waiting for an escort by the Pakistani army to cross the bridge and walk to their Promised Land, a land where Muslims could finally live free from the tyranny of the Hindu majority, free from the condescending colonialism of their white British masters. The public maidan, or open field, was overflowing with men, women, children, goats, cows, and bullock carts. Disease had started spreading due to lack of hygiene. People were hungry and thirsty. Some had trudged miles and miles from eastern Punjab. Hindus and Sikhs had relentlessly attacked their caravans throughout the journey, robbing, killing, and abducting their women. These victims of independence from the British had suffered unspeakable atrocities, but at last the end of their misery was nearby. They were close to the border between the newly carved nation states, India and Pakistan. The double-decker pedestrian and railway bridge on the Ravi was one mile from their encampment. One mile west, and they would be safe. But because of the fear of attacks, few had the courage to walk the final mile. People whiled away their time in misery and waited for the security of a promised army escort.

Water and food were running out. Their camp was filthy with excrement; children were crying of hunger; but the Muslim refugees did not dare step out of their camp. The atmosphere was poisonous. Mobs of Sikhs lay in wait. News from the west was escalating the passions of the Sikh and Hindu refugees. Refugees arriving from West Punjab were bringing stories of Muslim mobs butchering Hindu refugee caravans on their way to India at Narowal, a town ten miles west of Dera Baba Nanak. Stories of trains arriving from Lahore, awash in blood, packed with dead bodies, were circulating. People cleaning the bloody trains had discovered messages nailed to the severed arms of infants, instructing, "This is how you kill!"

The Punjabis were celebrating their newfound *azadi*—independence from the British—in their own unique way.

Indian army troops were seething in rage. They had watched helplessly as mobs assisted by the Pakistani army at Narowal train station, which was less than ten miles from the newly created border between India and Pakistan, had attacked and killed dozens of Hindus and Sikhs in cold blood. The protectors of the Hindus and

Sikhs, sent to escort them to India, were unable to prevent the massacre: there were only a few soldiers and a meager supply of ammunition. Now, in the safety of their territory, it was time for revenge.

The soldiers of the newly minted Indian army, purged of most of its Muslim troops, were making known their resolve to support Hindus and Sikhs in Dera Baba Nanak. As partisans, they were working up the refugees from West Punjab, who were chiefly Hindu. An appeal to base emotions was not at all difficult under the circumstances. "Look at what has happened to you. You were chased out of your homes like vermin. You watched Muslims butchering your children and raping your daughters and sisters. They have burned your homes down. They laugh at you, calling you spineless. Aren't you ashamed?" The manhood of the refugees was impugned; shame was promised to those who wouldn't participate in revenge; plans were made to even the score.

On August 29, 1947, when night fell, a murderous mob of Hindus and Sikhs was ready to strike. Armed with swords, spears, and daggers, they surrounded the Muslim refugee camp at the maidan. Those without weapons carried sticks to smash their enemies' heads. Farmers, shopkeepers, farm laborers, god-fearing simpletons, and loving family men turned into collaborators with the devil. Full of hatred, they attacked ruthlessly, killing men, women, and children indiscriminately. Many who tried to escape were shot dead by the soldiers. Terrified Muslims trying to flee ran directly into their killers. Many stampeded over their tormentors and sought refuge in the sugarcane fields or in culverts and gullies. They were hunted down and put to the sword. The massacre was thorough. Proud Punjabi Hindus had taken revenge against proud Punjabi Muslims. Refugees rendered destitute and homeless by circumstances beyond their control had murdered other destitute homeless refugees in revenge. The leaders of the partition movement, who were joyfully celebrating independence in Delhi and Karachi, had set in motion a vicious cycle of revenge. That evening, while Sikhs

and Hindus stabbed and sliced and troops fired on the defenseless, rain had come down in torrents. Floating in the streams, made
red with blood, were torn clothes, shoes, slippers, personal adorn-
ments, and the ripped bags and belongings of the Muslim refugees.

Early in the morning, someone had unlocked the door to let
the children out. Beeru stood in the courtyard, the ground beneath
his bare feet soggy. He was anxious to go out and explore with his
cousin Mohan, his best friend.

Beeru watched his cousin Hari swagger into sight. Hari was
dark like night and built like an ox. His lip-concealing thick mustache was waxed and tightly wound. Hari had defied death last
week and had managed to escape Kanjrur-Duttan. His two sisters
and mother were not so lucky; they had been killed. He threw a
cocky look at Beeru. "See what I have, a real gold ring. I took it
from a Muslim." Hari opened his palm, displaying a golden ring—
perhaps a wedding band.

"We taught them a lesson last night." Hari cackled.

"What happened? What lesson? Who did you teach a lesson?"

"Go and play with girls, you sissy. Quit bothering me, I have
better things to do than to talk to you," Hari snarled as he sauntered away.

Beeru and Mohan ran out of the dharmshala, their bare feet
digging into the ground.

Beeru's curiosity knew no bounds as he ran toward the Muslim refugee camp near the train station. His skinny legs outpaced
Mohan's. Suddenly he slipped on the muddy ground, stumbling
and falling facedown on a mud-smeared bundle. Mohan crashed
into Beeru and also took a tumble, rolling on moist earth, his
clothes and body covered with dark mud.

Beeru froze with fear as he stared at the bundle—it was a dead
body. Mohan pushed himself up and came rushing toward Beeru,
dragging him away from the dead body he was splayed on. Crouching on the ground, the two trembling boys held each other. Slowly

Beeru and Mohan stood up, their eyes fixed on a boy no older than
Mohan. The boy stared blankly at the clouds. His arms were spread
like the open wings of a large bat, his wounds concealed by a thick
layer of mud. Beeru cringed as he stared at the body—not Hindu
or Muslim, just a boy. No one could tell a Hindu from a Muslim
just by sight. The only indicators were how the person dressed and
whether he was circumcised. This was a child murdered simply for
the crime of being born in a Muslim family, by a murderer whose
crime was to be born in a Hindu family.

Who could kill a child? Did any of his uncles deliver the fatal
blow? Beeru wondered. Clutching each other's hands, Beeru and
Mohan walked toward the train tracks. Near the tracks, as far as
their eyes could see, the ground was littered with dead bodies. Vul-
tures hovered over the corpses, savoring the feast laid out for them.
Bodies lay in pieces. The victims' clothes were soaked in blood
mixed with mud. Severed heads lay on the ground, some facing the
ground, some staring at the sky. Arms, legs, hands, intestines, and
innards were strewn on the earth.

Beeru held Mohan's hand in a viselike grip. He looked at Mohan, imploring him with his eyes to return to the dharmshala.
Mohan seemed lost, his eyes glazed. They slowly stepped over a
body with its torso laid on the smooth and shiny train tracks, its
head resting on the beam of wood supporting the tracks. Beeru was
startled and jumped. Mohan was screaming in fear.

A naked man no older than Beeru's father was sprawled in front
of Mohan. A sword had slashed through his face below the upper
teeth, almost severing the lower half of his jaw. His lower teeth and
chin hung precariously. The tip of his tongue touched the ground,
separated from the palate, as if it were not even a part of the dead
man's mouth. His upper teeth and the roof of the mouth created a
ghastly sight. The dead man's severed leg rested a few feet from his
torso. Beeru involuntarily doubled up and began to retch.

Mohan had stopped screaming and was mumbling, "They are

coming; they are coming for us. Hurry, run, or they will kill us." Beeru couldn't see anyone coming but began running toward the dharmshala, stumbling over human remains, jumping over cadavers scattered on the ground, slipping and falling in mud. He kept running furiously.

———————

IN A LETTER SENT TO my paternal grandmother on September 15, 1947, my grandmother's sister Rani painted a picture of the dance of death she had faced while fleeing for her life across the new borders. Her words are stark and devoid of emotion:

Dear Sister, respectful greetings.

We have safely reached Ghaziabad. On the way, Mussalmans [Muslims] attacked us. We ran and it was difficult to save our lives but we survived. Kolo Masi's daughter-in-law went to Japuwal and probably Masi has gone to Benaras. Girdhari Lal and others are at Dere [Dera Baba Nanak] but Puran Chand and his entire family was killed. Mindi, Nimor, and Nargis were abducted by Mussalmans. Mindi's father was stabbed, he was lying wounded in Chandrake; don't know if he survived or not. Doctor Bakatram is dead. Bindo, Fufad (uncle), Ram Sharan, Bhimsen, Krishna, Bansidhar's wife, Shanti, Bansidhar's elder daughter-in-law were killed. Aunt Bishandei was killed. Rest is OK. Thank god that we arrived safely. Whatever [possessions] we saved in Lahore was left in Kanjrur. Love to children. Namaste to you and Jijaji.

Your sister, Rani

Rani and her family had fled from Muslim attacks in Lahore and escaped by going east to India. Her letter lists some of our family members she knew had been killed. It also reveals how our large family was displaced and scattered around India. My father lost contact with most of his relatives, never to meet them again. Clinging desperately to hope and fighting a battle with wrenching poverty, my grandfather managed to educate all of his children. None of them became a big landowner, but they managed to survive and make a life in the new land.

For some, the horrors never ended. Uncle Mohan lived a tortured life. He remained fearful of dark and empty rooms, refusing to enter them. He saw dead bodies everywhere; he would wake up in the middle of the night, screaming that people were trying to kill him. If the electricity went out (a routine occurrence in India) or if someone turned the lightbulb off, Mohan cried in terror. One day in 1982, he was walking into a grocery store when he was blown to bits by a bomb exploded by Khalistani terrorists in Punjab. People could not gather enough pieces of his body to cremate him.

Some relatives became persecutors. I met my granduncle Ziai in December of 2001. After five decades, my father had found him in Dehradun. Age had failed to diminish the ferocity that radiated from his copper-colored face. Chain-smoking, he spoke in a raspy and deep voice, reciting Urdu couplets from his poetry books. Ziai had watched his two sisters and uncles murdered while fleeing a ravaging Muslim mob in Kanjrur. He had seen Muslim neighbors turn thirsty for Hindu blood. To save his life, he jumped in and swam across a flooded Ravi.

When I asked my granduncle Ziai about the partition years later, he was unrepentant. After he escaped to the safety of India, he joined a band of young Hindus and Sikhs who were looting and killing Muslim refugees on their way to Pakistan. Ziai killed so many Muslim refugees in 1947 that he lost count. Ziai lit a cigarette and told me, "Son, times were like that, you can never understand. I was young and angry. I was roaming with a group of young

Punjabis, and the only thing on our mind was revenge. Whenever we killed a Muslim, that gave me great satisfaction!"

I tried to find remorse in the eyes of the man who was both a victim and a perpetrator of gruesome violence; I did not find any.

I asked him how many Muslims he had killed.

"I don't know, son, there were so many. I did not keep count." He rationalized: "Muslims were butchering entire trains filled with Hindu refugees; we were just retaliating." It almost sounded like a boast when he told me that I couldn't even imagine what he had been through. "I have walked on ground littered with dead bodies," he said.

I argued with him. "Our family members were killed by Muslims in Pakistan when they were desperately trying to escape, leaving behind their homes and ancestral lands. They did not even have time to gather their belongings to take with them. Do you think that the people you killed were also poor innocent refugees who were trying to escape to Pakistan?"

Ziai turned away from me, put his cigarette down, and paused. His eyes stared into nothingness. He slowly said, "Yes, those were also poor innocent people. I don't know how many innocent people I killed. There must have been some sinners amongst them, too."

I do not know what kept me at arm's length from the culture of violence of my Punjabi clan. Maybe it was metaphysical poetry, maybe it was spiritual music, maybe it was pure luck. While I took one lesson from our history, Raju took another.

4

A MAHARAJA
IN RAJA PARK

However much I become an expert
in breaking rocks;
If I live, there will always remain in the road,
another heavy stone.
— GHALIB

When I was a child, there were only three houses in our block. Raja Park fell outside the limits of Jaipur. No buses passed through our end of the town; to catch a three-wheeled motorized public taxi, a *tempo*, we had to walk to the end of Street No. 1. As I grew up, Raja Park also grew. Though the area was earmarked for Punjabi refugees, in the 1970s, non-refugees started to move into Raja Park. New houses were being constructed, wealthier people were moving in, auto rickshaws and buses supplemented the tempos, and a new residential development, Jawahar Nagar, came up across the shallow dry ravine, which marked the end of Raja Park and Jaipur. By the mid-1970s, Raja Park was not the desolate edge of the city.

It was in 1970 when a new neighbor quietly moved in two houses away from us. His name was Amar Singh Rathore; he exuded an aura of mystery. In our neighborhood, people introduced themselves to newcomers, visited and welcomed them, shared cups of tea, and talked for hours. Amar Singh was a recluse. People wondered who he was. Where were his wife and children? What was he doing in Raja Park, living in a large house all by himself? Why wouldn't he talk to anyone? Even his maid, who arrived early in the morning and left at dusk, avoided everyone. If he was hiding something or from someone, Amar Singh had picked the wrong place.

A few months after he made Raja Park his home, Amar Singh turned into a celebrity. He became the king of our street, royalty among the Punjabi refugees. It happened after we caught sight of his finely dressed relatives visiting him in a 1960s Chevrolet Impala. No one in Raja Park had a car. During the 1960s, India manufactured only some thirty thousand cars a year. Most people rode on bicycles; a privileged few owned motorized scooters—which required a deposit and a wait of several years. In the 1970s, very few Indians could afford a car. Those who could had a choice of two boxy models: a stolid Ambassador, based on outmoded Morris Oxford Series III cars, or an equally dull ancient Fiat model. To purchase either one, one had to put down a deposit and then wait for years. An American model was a true rarity and astronomically expensive.

Amar Singh was the second son of Maharaja Sadul Singh, the maharaja and the last ruler of the princely state of Bikaner. As a prince, he grew up in opulence in the Lalgarh Palace, Bikaner. He was adopted by the estate of Maharaja Sadul Singh's younger brother, Prince Bijey Singh, after the mysterious death of Bijey Singh in 1932. Bijey Singh was being groomed to succeed the throne by Maharaja Ganga Singh. The *New York Times* called the prince's death an accident when the twenty-nine-year-old heir apparent shot himself while examining an automatic pistol. It is said that Prince Bijey was having an affair with his aide-de-camp Hem Singh. Homosexuality is taboo in India; in a masochistic Rajput warrior culture, it is fatally sacrilegious. When the affair came to light, both of them committed suicide. After his adoption, Amar Singh became the heir to Prince Bijey Singh's estate and the revenue derived from it. Amar Singh had been endowed with great wealth, including land grants and his part of the royal fortunes of Bikaner.

While people gossiped about trunks full of money Amar Singh had in his house, I was impressed by one particular fact: Amar Singh owned a cinema hall, Ganga Talkies (later renamed Minerva), in one of the busiest areas in downtown Jaipur. As far as I was concerned, no one could be richer than the owner of a movie theater.

The neighbors eventually found out Amar Singh's connections with royalty. Everyone started fawning all over him. Any time he stepped out of his house, people tried to engage him in conversation. They competed to be acknowledged by the Raja Sahib, as everyone now addressed him, inviting him over for visits. Amar Singh ignored the toadying, though he remained unassuming and respectful. With time, he did become slightly sociable. He would sometimes greet the neighbors and engage in small talk. Later, Amar Singh's elder son Chandrashekhar Rathore moved in with his father.

———

AFTER PARTITION, THE INDIAN GOVERNMENT gave my grandfather, Bihari Lal, whom we addressed as Bhapaji, a few acres of agricultural land in Punjab, in compensation for the lands stolen from our family in Kanjrur. It was a pittance, but it tided him over. Though his family lived in Jaipur, Bhapaji was a bank employee about 150 miles away in Agra. Whenever he could slip away from work, he would go and supervise the hired landless farmers. He made certain that the poor cultivators did not cheat him out of his share. His greed and suspicious nature lead him to spend his vacation time and holidays in the village. Thus he rarely spent any time with his family. On his rare visits to Jaipur, which happened once or maybe twice a year, he rarely stayed for more than a couple of days.

He was dark, bald, and squat. His fleshy body stood in contrast to his thin-as-a-stick family members. Bhapaji was loutish, loud, and embarrassingly stingy. He never wished to spend a dime on anything. He argued, bargained, and quarreled with the vendors. He fought over the fare—mere pennies—with sickly looking rickshaw pullers. He never gave gifts or purchased anything for the family.

He never did read a book, listen to music, visit anyone, or go to the movies. Seemingly uninterested in anything except eating, he whiled away his days sitting on a wicker chair in the front yard. He rarely spent time with my grandmother or talked to her, unless he wanted food; they slept in different rooms. Like all the patriarchs of his generation, he brooked no disagreement; he blew up if somebody questioned him.

Bhapaji ignored Raju and me; he rarely talked to us except to issue some gratuitously condescending remark. I did not know what to make of my grandfather. I avoided him at any cost. I do not recall ever having a conversation with him. He did not acknowledge my presence, except when he felt like making fun of me. The day I left home, never to see him again, I did not say goodbye. In his entire life, he never once uttered a kind word to me. The reason for his disdain was simple: we were his enemies! Bhapaji hated our father, VK.

There had been a split in our home. Bhapaji, my youngest uncle Avinash, Bhabhiji (my grandmother), and my aunts Nanda and Kanta, belonged to one faction. My second uncle, SK, kind, sensitive, and cowardly, followed the Avinash group; his was the path of least resistance for a person who avoided arguments and wished to take no position. My father, mother, Raju, and I made up the opposing faction.

The source of conflict was the house we lived in. Bhapaji and VK each claimed to be the legitimate owner of the house. VK owned the deed to the house and insisted that with Surinder, he had paid for the construction of the house over the course of a decade. Bhapaji claimed that he had contributed more toward the construction of the house.

It was around 1970 or 1971 when Bhapaji decided he wanted the deed of the house in his name. Early one morning, Bhapaji had summoned VK to his room. The conversation was short; within a few minutes, a deafening roar shattered the calm of the morning. It was Bhapaji screaming at the top of his lungs. Cursing out loud, he stomped out of his room, rushing to the storage room. My father emerged from Bhapaji's room. I don't know what was more disturbing: Bhapaji's screaming at my father, calling him a bastard, a son of a bitch, a motherfucker, a sisterfucker, and other choice swear words; or his threatening to break our limbs and throw us out of the house. Bhapaji dragged our belongings from the storage room and started throwing trunks, suitcases, and bags in the courtyard. If I required proof that anger made one stronger, I was looking at it; the old man was grabbing, lifting, and throwing heavy objects with ease and fluidity, as if he was tossing feathers.

I stood next to our room, wondering whether Bhapaji would follow up on his threats to shatter our anklebones and kick us out of the house. VK stood stoically and watched the pile of metal trunks and cheap faux leather suitcases grow taller. He watched quietly as Bhapaji tossed the last of our bags in the courtyard. The din of metal trunks hitting the ground finally came to a stop, though Bhapaji's cursing continued. No one else uttered a word.

After Bhapaji stomped back to his room, VK haphazardly began picking up the widely scattered trunks, suitcases, and bags, placing them next to the living room wall.

Bhapaji was furious because my father had refused to sign away the deed of the house. The day our belongings were tossed out in the courtyard was the day our family split apart. From that day on, we lived in the same house, shared the same bathroom and shower, inhabited adjacent rooms, but were enemies. The air was thick with hate, anger, and frustration. And Avinash was the angriest and most frustrated of all. Why wouldn't my puny father capitulate to Bhapaji's intimidation and sign away the property?

———

BLOWS KEPT LANDING ON MY back, legs, and head. The attack was sudden and disabling. Avinash, my father's youngest brother, was mercilessly hitting me with a cane. He had recently broken his leg in an accident and needed the cane to walk; the cane came in handy for beating a seven-year-old. Each time the cane struck, I flailed about and screamed. I was desperately biting my lips, trying not to cry, but I was wailing and writhing in the Raja Park sand. Getting beaten up and being humiliated in public was agonizing enough. It was distressing to see Happy and Bawa, my playmates, standing across the narrow street, smirking and enjoying the show.

After a miserably painful beating that seemed to last forever, Avinash grabbed my arm, pulled me up like a rag doll, and stomped away. He was angry because I was running in the street, chasing after a kite. I guess it was a serious offense; only low-class kids were supposed to do this. As I stood up, I saw Avinash's wife and my grandmother. They had been enjoying the show. There was no pity in their eyes. Raju was standing a few yards away. He had been chasing the kites, too, but survived the beating.

When my parents were not around, Avinash, Nanda, and

Kanta often harassed and abused us. We were the targets of cold hard stares, abuse over petty things, and merciless thrashings. I was seven, Raju was four years older. Avinash was always searching for an excuse to humiliate and beat me, though occasionally he didn't even bother to come up with an excuse.

Later, sitting cautiously on my cot, I felt my chest and back for bruising. I was deeply ashamed: I had been beaten right in the middle of the street, in front of my neighbors; my two playmates had watched from across the street. I was worried that they would think less of me and refuse to play with me. The beating had to be my own fault.

Avinash, in his early thirties, was a bald man with a shiny dome, a round face, and a thick mustache. Not only did he look like a thug, he took pride in acting like one. He was boorish and loud, often bragging about how he had bullied a rickshaw puller or a pushcart vendor or how he had scared someone. After finishing college, he had joined the Indian army as a veterinary doctor. After a short stint in the army, he had returned to Jaipur. He ruined my early childhood and terrorized Raju and me when we were growing up. Although I was a regular victim of his verbal and physical abuse, I clearly understood that Avinash was not after me. Raju and I were stand-ins for our father. He was trying to make VK and our lives miserable, hoping to force VK into signing away the deed to the house. But the plan did not work. VK did not sign away the house to Bhapaji. Though VK was smaller, weaker, and outnumbered, he refused to be intimidated or to yield. The war continued; Raju and I were pawns caught in the middle, regularly beaten, humiliated, and abused.

I WAS READING INSIDE OUR room when I heard someone approach; it was Raju. Grinning broadly, he grabbed my arm and began to try to drag me out. "Come, Rathore uncle wants to see you," he said excitedly. I could not fathom how Raju had made the acquaintance of Amar Singh, whom we called Rathore uncle and others addressed as Raja Sahib. Why would Raja Sahib be interested in meeting a nonperson like me?

"I am not going." I was emphatic. I was shy and awkward with people and had trouble making friends; the thought of Amar Singh's imposing frame and bearing further unnerved me.

"Come on, he is very nice; he will give you Coca-Cola," Raju wheedled, but the enticement was not effective.

"I am certainly not going." I tensed up and tried to break Raju's grip. Only rich people drank Coke; sugary milk tea was the customary drink of the common man in Raja Park. I had never tasted Coke, and as Raju attempted to drag me out, I had already decided that I did not like it anyway.

Raju's persuasions were not working. "Why don't you go and say namaste to Raja Sahib?" my father suggested, breaking the deadlock. Now it was two against one; in any case, I could not refuse Father's suggestion. Reluctantly, dragging my feet, I followed Raju, only to find myself in front of the iron gate to Amar Singh's house. The gate led to the driveway, which ran the length of the house and ended at the wall of the next house. As one entered from the gate, Amar Singh's house was to the right of the driveway. To the left was a lush lawn bordered by rows of colorful flowering plants. Containers potted with ornamentals lined the driveway. Raja Sahib sat on a wicker chair in the middle of the lawn, his large girth making the seat a tight fit; I wondered if he had to struggle to get out of the chair. Chandrashekhar sat opposite Amar Singh. Both were large and imposing men. I still remember being awed by these bear-like larger-than-life humans. They were staring at me with gentle and amused smiles.

Amar Singh greeted me. Though I was too ill at ease to hear

anything he said, I was struck by his voice. He had a sweet, high-pitched singsong voice, which didn't seem to match his massive frame. I folded my hand and feebly uttered namaste. "Sunil, would you like a cold drink? Some Coca-Cola?" Chandrashekhar asked. He also spoke in a soft, melodious high-pitched voice, just like his father's. Their manner of speech was a pleasant contrast to my family's loud, boorish way of talking.

"No, no, no. I don't want it," I protested.

I was wondering why these privileged high-class individuals were being so kind to a shabbily dressed child who clearly did not belong in their ranks. Something was wrong!

"Don't be shy. Why don't you want a cold drink?" I stared at the ground as Chandrashekhar asked me in his sweet singsong again. I shook my head to refuse the drink. Raju gladly accepted the offer. All I remember of the visit was being self-conscious and uneasy as Amar Singh and Chandrashekhar tried to make small talk with Raju and me. I walked back from Amar Singh's house wondering why the Raja Sahib wanted to meet me.

———

SOMEHOW THE FORTY-SIX-YEAR-OLD RAJA SAHIB, scion of an illustrious warrior Rajput clan, hit it off with an eleven-year-old Punjabi refugee. Amar Singh developed a special fondness for Raju. Of all the neighbors, only Raju and I were permitted in Amar Singh's house. Raju became a regular visitor and spent hours with Amar Singh. Sometimes I accompanied Raju; other times I went by myself. In winters, we sat on the terrace of Amar Singh's house, soaking in the sun, looking at the people passing by, and listening to the fascinating stories he told. The combination of warm sunshine and extraordinary tales made time fly. As he drank cup after cup of tea, and as his maid continued to bring fresh pots, Amar Singh unabashedly regaled us with stories grown-ups would generally not share with children. Enthralled, we listened in awe as he talked about his World War II tour, his trips to foreign countries, and, rarely, his life in Bikaner. We always wanted to hear about battles and wars, but instead of telling us about soldiers, battles, gunfights, or camel regiments, he amused and sometimes embarrassed me with lurid tales from the front. Only later on did I learn that most of the Indian princes and kings did not see any action on the front. The British empire didn't want their loyalist rulers in danger. The kings and princes proudly inspected the soldiers they had conscripted for the war effort, received fancy medals, and returned with army ranks they did not earn or deserve. And the royals at the front enjoyed wild parties and shows.

Women and fun were never lacking in Amar Singh's stories. He found white women a nuisance, because they could never have enough kisses and kept asking to be kissed during sex. Chuckling, Amar Singh painted vivid pictures of the shows he had attended where showgirls tried to have sex with a donkey, among other raunchy acts. "She kept sliding back; I don't think the donkey succeeded," Amar Singh said skeptically. I was mortified, not knowing how to react to such stories. Raju, on the other hand, laughed gleefully.

Amar Singh presented his World War II tour as fun-filled. His observations from the kingdom of Bikaner were equally lurid, but he and Raju took great delight in them. I winced when he described how the subjects who failed to pay their taxes had dried pellets of camel dung hammered in their anus.

Though Amar Singh told many stories, he never talked about his family, his relatives, or his royal heritage or upbringing. Amar Singh's wife lived separately in Japan, and his son of high school age lived in Bombay. His two married daughters rarely visited him. Even as naive as I was, I knew something was wrong between Amar

Singh and his family. His relatives scarcely came to visit. Not too long after Chandrashekhar came to live with Amar Singh, they had some differences and Chandrashekhar moved out of his father's home to a place a few blocks away. Later he moved away from Raja Park entirely. I benefited from this move, as Amar Singh's younger son, Rajesh Rathore, moved in with his father.

Rajesh was five years older than me. We spent much time together, making a rather unlikely duo of a scrawny Punjabi refugee boy and a chubby Rajput prince. It was hard to miss the vast gulf between us or to ignore the extreme difference in our social rank. Rajesh ate better, had closets full of fancy clothes and shoes, went to movies whenever he wanted to, spoke fluent English, and never had a worry; he did not even go to college. One house away, I had none of these advantages. I was so ashamed of our penurious shared joint family house that despite our friendship, I never invited Rajesh for a visit. However, despite all our dissimilarities, Rajesh spent all the time he could with me.

After school, I visited Rajesh almost daily. He was obsessed with movies and Hindi film music. We listened to music for hours on his fancy stereo system. Rajesh also loved American comic books; I was introduced to stacks of *Archie* comics, which he let me borrow. Above all, Rajesh took me to the movies and paid for my tickets. Besides paying for my tickets, he took me to Ram Niwas Bagh, an expansive garden adjacent to downtown, and treated me to snacks and pineapple juice. Raju had become a good friend of Amar Singh, I of Rajesh.

I found it comforting to visit Amar Singh's quiet and spacious house; it was a delightful escape from our dreary home. I am certain that Raju's experience was the same. The tranquility of Amar Singh's home, compared to the chaos of ours, was refreshing. It taught me there was kindness and care in the world, even though it existed outside our home. I learned that humiliating others and abusing children was not normal. The biggest lesson was that a little respect and kindness, sharing a little humanity, could transform the lives of others. I found out that even some of the rich could be down to earth. That sharing of humanity worked in Raja Park; it also worked in the crime-ridden streets of Los Angeles, when some hardened criminals ended up dropping their facade of bravado and anger in front of me. Why didn't this sharing of humanity work with Raju? He was treated in an extra special manner by Amar Singh.

5

A JAIPUR ROMANCE

O Lord, with whom shall I share
The pain of separation
The fire of love burns and smolders within me
Wandering in the vast desert and wilderness
I have not yet found my beloved
Sorrow is the food I eat, heartache my drink
My breathing is reduced to hot sighs
I wander looking for my beloved,
Yet my beloved remains within me
Hussain the Fakir says
Only union with the beloved will bring me joy.
—MADHO LAL HUSSAIN

Jaipur, India, 1981–82

Hindi movies have bewitched the Indian people since the early part of the twentieth century. Once the silent films became talkies, their popularity and hold on the public increased. Romantic Bollywood tragedies with their haunting love songs captivated the Indian masses. Creating an alternative reality, cinema portrayed ideal love in an orthodox society where love between a man and woman was mostly a fantasy; romance and love was something you read about in novels. In a land where arranged marriages are still the norm, a favorite storyline of Hindi cinema is young adults falling madly in love.

Outside the cinema, in real life, the blossoming of love had to wait until after the arranged marriage. Family elders searched for eligible brides and grooms within their own clans. Suitable matches were based upon caste, social rank, and the negotiation of a suitable dowry; consent of the potential bride and groom was not a prerequisite. Once the two strangers were brought together and married in an elaborate ceremony lasting for days, they might eventually find themselves in love. But as Ghalib says, "You can't force love, it's a fire; which doesn't light if you want; and doesn't extinguish if you try." Marriages without love were a reality no one wished to confront. Even if the marriages did not work, couples went through their lives, had children, and separated only at death. Divorce was unthinkable; there was no escape from misery or abuse, especially for women.

While it may have been next to impossible to romance someone before marriage, for an investment of two to four rupees, one could watch stars of the screen fall in love and feel the magic of romance vicariously. The most popular Hindi romantic movies were also the most tragic. Love always failed; the lovers were always separated; hearts always broke. The underlying message was that true love was destined to fail; love was nothing but eternal suffering. *Devdas* is one of the greatest romantic movies ever to come from Bollywood. The film had such powerful impact that across India, the word *Devdas*, which literally means servant of God, is understood to mean a broken-hearted male lover. Based on a 1917 Bengali romance novel by Sharat Chandra Chattopadhyay, *Devdas* was first produced in 1928; many notable film versions followed. If someone brought the essence of Devdas to the screen, it was Dilip Kumar, in the version directed by the Vittorio De Sica of the Indian film industry: Bimal Roy.

Devdas is a heart-rending story of thwarted love. Devdas comes from a wealthy family; his childhood friend Parvati (Paro) comes from a lower social rank. Due to misunderstandings, family pride, and class conflict, Devdas is unable to marry Paro. Her family marries Paro off to a widower. Unable to stop Paro's marriage, Devdas promises to visit her before he dies. Separated from Paro, the lovelorn Devdas sinks into depression and drowns his sorrow in alcohol. The combination of excessive drinking and despair is Devdas's path to a slow agonizing death—a drawn-out suicide. As Devdas's health deteriorates and he senses the approach of his death, he sets out to fulfill his vow to Paro. He travels to her village but ends up dying on her doorstep without meeting her. In the last scene of the film, after learning of his death, a distressed Paro runs toward the door for a final look at Devdas. Before she can exit the gate of her house, her family stops her. The two lovers fail to unite in life and in death.

As a young boy, even though I knew that Hindi movies portrayed a fictional world, I believed them nevertheless. I was convinced that love ended in tragedy. I wondered why anyone would fall in love when union with the beloved was impossible. In the India I lived in, there were no love stories, no love marriages, and no overt romance or tenderness—even between married couples. Couples did not hold hands, did not embrace, did not kiss; all of that was saved for moments of privacy—which were rare in joint family households. Forget love—one couldn't even look at the opposite sex. Rajasthan was a conservative state. Boys and girls were not allowed to mix; gender segregation was strictly enforced. In my co-ed school, boys and girls sat in different rows, eyes fixed on the blackboard in the front; we didn't stare at each other or play together; we ate separately. The strict wall of separation between girls and boys, however, could not kill teenage fantasies. One could always dream or steal a glance or fall in love—quietly and secretly.

———————

I WAS IN TENTH GRADE when I first fell in love. Shabnam, the object of my affections, was seventeen; I was fifteen. Because Shabnam was tall, she stood out among the rest of the girls. She had raven-black wavy hair and a fair face. The shapeless school uniform failed to hide her curvaceous body. She carried herself with confidence.

The contrast between us couldn't have been more obvious. She was the noble Heer of the epic Punjabi romance *Heer Ranjha*— driven to the school in a chauffeured car. I pedaled a rusted, creaky bicycle three miles to school. Shabnam was beautiful and poised; I was insecure and awkward. But as Ghalib said, who can control love?

It is said that love overcomes adversity, gives power to the weak, lifts the fallen, brings smiles to sad faces, and moistens the eyes with tears of joy. A popular Hindi song goes, "Whether you listen or not, I will tell you the condition of my heart" (*hum hale dil sunay-enge, suniye ke na suniye*). Of course, the other party cannot listen to you unless you speak first. I did not have the courage to walk up to Shabnam and say hello. I lost the power of speech every time I saw her. I consoled myself with the poetry of love, which somehow always aptly described my lovesickness. For instance, Ghalib had said:

My gift of speech is gone, but even if I still had it,
What reason would I have to put desire into words?

How could I put my desire into words? I had nothing to offer this noble beauty! And if I did manage to speak, would Shabnam pay any attention to my words? Love had not given me strength or courage, and it didn't overcome any obstacles. I sighed, I pined, and I dreamed. I was too shy, too realistic about my chances, too cowardly; I was secretly in love. The love *had* to remain secret! Who could help me? Could I ask my friends to please go and tell Shabnam I am madly in love with her and I would like her to be mine?

I had shared my secret with Amar Singh a few times. He would listen quietly without any comment. One day when I was particularly sad, he said, "Why don't you tell her? Then we will get the band to play for you." In India, a loud band, playing popular film songs, leads the groom's marriage procession toward the bride's residence.

I was taken aback by Amar Singh's flippant remark. "What? What do you mean? Is it that simple? I don't have a job, I am in tenth grade, I live in a small room with my brother!"

"Why fall in love when you cannot marry her?" Amar Singh retorted. "What's the point of looking from a distance? If you are not determined to get the girl, then don't say you love her."

Amar Singh's harsh words hurt but were a wake-up call. I *thought* I was determined to get the girl, but in reality, I was unable and unwilling to take the next step. My heart was foolish and irrational, yet logical at the same time; it told me that I had no chance while it longed for Shabnam. Every day between my classes, I walked to Shabnam's classroom, hoping to catch a glimpse of her. During the lunch break, I followed her at a distance and watched her eat with friends. I biked miles to hover around her home, just to catch a sight. Months passed, but she never found out about my mad passion for her.

Shabnam graduated from high school and left for college. My dreams were shattered. In the Hindi tragedies, poor boy and rich girl (or vice versa) meet accidentally and fall in love. The happy lovers run around trees, crooning love songs, pledging to be together forever. But then heartless parents and families come between them; the lovers are cruelly separated and love is thwarted.

In my case, we "lovers" were cruelly separated without ever being together.

The next time I fell in love, I ignored Amar Singh's advice once again.

———————

MUNNA LAL WAS A FRIEND of Raju's. He worked for the Rajasthan police as the lowest-ranking cop. He was a 400-meter sprinter. Instead of patrolling the city, he trained twice a day and represented his employer in track and field events. Slim, tall, and athletic, he was outgoing, friendly, and kind. I do not remember how I met him, but he somehow convinced me to become a runner. He introduced me to the Sawai Man Singh Stadium (SMS), Jaipur's main training ground for professional athletes, and to the contingent of runners representing different agencies. By the time I was in high school, I was competing in 5K races. Munna Lal and I became running partners until I gravitated toward the boxing team at the SMS. Our friendship continued.

In many ways, Munna Lal was responsible for my becoming a man—and finding that rare treasure which many seek but never find: love. Of all places, I found it in a sweaty, hot, and unromantic place. One evening I was working out with free weights in the stadium. The gardeners were watering the grassy area in the middle of the cinder track. The women's field hockey team was practicing on one side, the gymnasts training on the other. I was doing clean and jerk lifts, sweating and struggling to push the bar above my head. After the set, I bent down to place the barbell on the mat. As I straightened, a pair of legs appeared in my peripheral vision. These legs were different from any I had ever seen before. The soft, smooth, round curvy calves were toned, almost muscular, and yet feminine. As I stood up, my gaze moved from the calves to shapely milky white thighs to the face. I was looking at a young blond girl with crystal-blue eyes. Her face was freckled, giving contrast to her fair skin. Her pink lips reminded me of Mughal-era poet Momin's couplet: *How elegant are her delicate lips, they put rose petals to shame.*

It was the first time in my life I had looked directly into the eyes of a young woman. We smiled at each other.

I wondered how she ended up inside the stadium. Sportswomen generally accompanied other women or a male family member to the stadium. Although India had a woman prime minister and many powerful women leaders, being a woman here was not easy. India was and still remains a male-dominated society. Male chauvinism permeates every rung of society.

A woman walking alone is an easy target of harassment, groping, and catcalls. The famous Bollywood cinema perpetuated sexism and the dehumanization of women; the courtship of the heroines in Hindi movies was nothing less than harassment, sexual battery, and assault. Women reporting rapes were treated as criminals and considered dishonorable. No wonder most sexual assaults are never reported in India.

The crowd in the stadium was even less enlightened than the average Indian male. Many of the athletes came from villages with strict authoritarian and sexist attitudes toward women. While most were helpful and respectful, some were quite capable and willing to molest a woman if they got a chance. A naive white teenager was at risk, especially since white women in India had, and still have, a reputation of looseness and promiscuity. The prevailing attitude was that because Western women dressed the way they did, they were asking for it. I wanted to tell this teenager to be careful of the men at the stadium, that it was risky for her to be here by herself; but how does a male stranger tell a female stranger to be wary of all other males?

She was looking at me inquisitively. "Hello! How are you?" I smiled. "Are you looking for someone?" I spoke passable conversational English with a heavy Indian accent.

"Hello!" She smiled back. "I injured my left foot. I want to try some exercises with weights to strengthen my feet and legs," she said.

"Sure, we have the equipment right here. I will show you some exercises using free weights."

A lissome young blonde in dark blue running shorts and a thin white cotton T-shirt stood next to me; her blue eyes and her alluring thighs kept drawing me in. One does not see the thighs of females in India; Indian women do not wear shorts. I was straining to act normal and not gawk at her.

I showed her different strengthening exercises. Standing within inches of her, I felt self-conscious handing over weights to her and showing her the proper form while lifting. I was drenched in sweat, with sweat pouring into my eyes; I probably smelled awful, but she smelled good.

Her name was Wes, and she had recently arrived from the United States with her adoptive father and his family. Her adoptive father, David Ray, a visiting professor at the local university, had invited her to India. Her family, including her stepmother, Judy, and her stepsister, Sapphina, was staying at the university guesthouse and searching for accommodations. They were going to spend a year in India. Wes was a runner. Someone at the guesthouse had recommended SMS to her for training.

Wes was friendly, outgoing, self-assured, and warm. She asked me questions about myself and my family. By the time the workout ended, I felt comfortable with her; however, not comfortable enough to warn her about being careful around men at the stadium. After the workout, I walked her to her shiny new bicycle; its black frame had WES painted on it in vivid white letters. "I come to train here at five in the morning and also at five in the evening," I offered. "If you come here in the evening, I can help you with the weights."

"Sure, I will come, but I can't come tomorrow, it will have to be the day after," she said. I resolved to diplomatically advise her the next time I saw her that many guys in the stadium posed a threat to her. She should be very careful and always travel with someone she could trust; being alone was dangerous. I was relieved that Wes had not run straight into one of the wicked ones that day; some of them even believed that a white woman would never complain if they forced themselves upon her.

"Great! I will wait for you next to the track." I watched her climb on her bike and take off. She looked back and waved at me. I waved back with a big smile.

Within seconds after she left, I was mobbed. I was so engrossed with Wes that I had forgotten all the other athletes around us. They were showering me with questions and teasing me good-naturedly, as well as pestering me. "Dutta, are we so inferior that you won't even look at us?" "Wow, you two were smiling and talking like lovers." "What did she say, what did she say?" "*Ladki pat gayi kya?*" (meaning "Did you win her over?") "When is she giving her pussy to you?" "Are you going to keep her all for yourself? Why don't you give us a taste, too?"

"What are you talking about? I was just showing her some exercises," I protested.

Some of the pestering was men's locker-room talk; some was vile. "Don't fool us, we know everything. You were trying to get her in bed." "She will give it to you, she gives it to everyone. What's wrong if you take a dip?" I was disgusted by the innuendos. "How do you know what she does? How can you malign her like this?" I challenged.

"White women fuck with everyone. You are just another dick. She will squeeze you like a lemon and discard you!"

Incensed, I walked away from the group. I'd had a pleasant chat with a friendly girl; these unenlightened provincial boors had soured my experience.

As I biked back home, the angry sneer on my face faded. Thinking of her delightful company, I burst out in a smile. I had enjoyed a lovely time with a charming girl and I was going to see her again in two days. Why pay attention to those crude imbeciles? They were just venting their jealousy.

Two days later, I arrived half an hour early to the stadium. I was not going to miss meeting Wes. At five P.M., the rest of the runners began their warm-up, running laps on the cinder track. I quietly disappeared. She hadn't arrived yet. I was unwilling to risk her not finding me and instead running into some thickheaded lout. I sat in the grass watching the east entrance of the stadium, getting more anxious by the minute. What had happened? She had said she would be here by five! Did she come in through the other entrance I was not watching? Maybe she didn't see me; maybe she looked for me and then left.

"What are you doing sitting here? We are done!" I was focusing on the east gate and had not noticed Mahindra, who was standing over me. Startled, I mumbled, "I am just, I am just . . ." Mahindra cut me off with a knowing look. "You are looking for that white girl!" "No, no, I am not," I stammered unconvincingly. Mahindra smiled at me as he walked away. Mahindra, an Indian Railways athlete, was a kindhearted Sikh from Punjab. Before some other acquaintance cornered and badgered me for sitting out the training and staring at the entrance gate for an hour, I sneaked out of the stadium.

I felt crestfallen. I had been so looking forward to seeing her. Would she come tomorrow? Maybe she was busy and will show up tomorrow, I hoped.

Every day for the next several days, I waited for her and kept my eyes peeled. I looked at the gates every few minutes while running, stretching, or having a conversation with others. She didn't come back.

———

AMAR SINGH'S YOUNGEST SON, Rajesh Rathore, had recently returned from visiting his mother in Japan. He had brought bags full of expensive gifts, which he was only too happy to show off to me. He let me borrow a bright white Adidas tracksuit with blue stripes. It was something I could never afford; I immediately decided to flaunt it by going on a long run. I had run about three kilometers and turned toward the university, heading toward the road going east from the university guesthouse. The road ended up in deep sand dunes, which tested the endurance of the best runners; this was my favorite place to run.

The road around the university campus had no sidewalk. Running at the edge of the road, I was dodging buses, bicycles, rickshaws, pedestrians, stray dogs, broken glass scattered on the road, and groups of students milling about. My eyes were drawn to an approaching bicycle. The blond rider stood out in the mass of humanity around her. I slowed my pace, gazing at her. She did not notice me at first. I waved at her, and there was a flicker of recognition in her eyes. She continued to pedal and I began to jog beside her. "Why didn't you come to the stadium? I was waiting for you." She hadn't come to the stadium as her foot had been hurting, she explained. It was hard to carry on a conversation while keeping pace with her bicycle and both of us dodging the chaotic traffic.

I was overjoyed at this second chance encounter with Wes. I don't remember if she promised to return to the stadium and train with me; if she was not going to the stadium, I didn't have to warn her about being safe. After a mile or so, we waved goodbye and I jogged back home.

MY MOTHER WORKED AT THE university library. From childhood, I had spent numerous hours in the quiet, still, and musty bookstacks, poring over books of poetry, philosophy, literature, fiction. Reading was an obsession. Reading kept me occupied; it was free entertainment for a permanently cash-strapped boy. Before I was in high school, I had read Kafka, most of the works of the Greek philosophers, Dickens, Voltaire, Milton, Wordsworth, Mark Twain, and others; I had read almost all the ancient Indian religious texts, which caused me to turn into an atheist. I also developed a great love for Urdu poetry. I went to the library almost every day, taking shortcuts through the sand dunes and the dry, barren open land that separated the campus buildings.

In the summer of 1981, I had finished high school and hoped to go to an agricultural university in Hisar, Haryana. As an idealist, I wanted to help Indian farmers, a vast majority of whom lived a precarious and impoverished life, eking out a livelihood from tiny parcels of land. Due to a bureaucratic hitch, my college application was not accepted in Hisar. In fall I started college in Jaipur as a biology major. I also enrolled in an evening basic German class.

As I walked to my first German class, the university was quiet, walkways and buildings emptied of the throngs of daytime students. German classes were held in a thick-walled sandstone building that looked abandoned. I walked the hallways looking for my class; it turned out to be the sole room in use at the time. The instructor, Dr. Pawan Surana, turned out to be a short, lively Rajasthani woman with an engaging manner.

After an hour, the class was dismissed. A gaggle of students was waiting outside for their class in advanced-level German. Clutching my books, I walked down the stairs. As I stepped out of the building, a bicycle sped up. A harried-looking Wes, late for the class, jumped off her bike and pulled some books from the bike bag. In three weeks, we had come across each other a third time. She was wearing a thin grayish-green shirt and a colorful long skirt. She quickly greeted me and rushed off to her class.

One moment in our life can transform our world. Ignorant, living in the middle of that moment, we rarely notice its significance. As Bedil, the great seventeenth-century Sufi saint and poet from Delhi, says:

Waves and foam are oblivious
to the depth of the ocean.
Unaware, incompetent, and full of conceit,
We search for secrets of reality.

I, too, was completely oblivious at the time. How was I to know
that a chance meeting with an American teenager would become
a life-changing event? In retrospect, it seems that what happens
is meant to happen. At that point I did not believe in destiny. I
thought we made our choices and aimed for our goals. But is that
the truth? Or was Urdu poet Mir correct when he declared:

Why we helpless are accused of free will?
You do as You please, we are needlessly blamed!

In hindsight, I see all life-transforming events—love, enmity,
friendships, tragedies—as predetermined. I still don't believe in
destiny; things are governed by mathematical probability and just
happen. We claim that what happened to us was an outcome of
our efforts, that we had a choice. But then, doesn't probability also
become destiny?

A few days later, after German class, I invited Wes to visit my
home. To my surprise, she agreed. Both my offer and her acceptance
were unusual. Ashamed of my dysfunctional family and the overrid-
ing hostility in our house, I had been embarrassed to invite anyone
for a visit. I had known Rajesh for over three years, but though he
lived only a few steps away from me, he had not been to our place.

On the day of the visit, I waited for that knock on the door, but
I was also aware that she had forgotten her promise at the stadium.
Maybe she would come, maybe she wouldn't. But that afternoon,
while my parents were at work and Raju was at college, a young
American blonde on a bicycle appeared on Street No. 6, stopping
in front of our house.

From the front window, I saw her open the gate. She walked under the grapevine covering the driveway and knocked at the door. I stood near the door, shy, conflicted, and delighted. She stepped inside the small area that served as our living room as well as my parents' bedroom, with a sofa up against one wall and a bed against the other. I was thrilled, I was embarrassed: a privileged American girl had stepped in the house of an underprivileged Indian boy. *She has come to my house, it's God's blessing; I look at her, and then I look at the condition of my house!* I wondered if she could read the anxiety writ large on my face. She had showed up. What was I supposed to do? I had nothing to show her, nothing to entertain her with, no gadgets to impress her with. I was not witty, amusing, or outgoing. What the heck was I thinking when I invited her? What was I going to do?

"Is this your room?" she asked. I stirred out of my stupor. "No, this is my parents' room; my brother and I live in the next room."

Hesitant and self-conscious, I offered her a cup of tea. "Sure," she said—unlike a typical Indian visitor, who must refuse tea or snacks several times before saying yes.

"Let me make you some tea." I walked back through my room into the small kitchen. As I looked for a pan to make chai, the traditional Indian tea, I was startled to find her standing right behind me. She had followed me into the kitchen.

Wes asked me to explain what I was doing while I added cardamom, ginger, and milk to water and boiled it with tea leaves. In traditional Indian hospitality, the host is obliged to make the guest comfortable; instead she was the one putting me at ease.

Wes turned out to be a good-hearted down-to-earth person. And unlike me, she loved to talk. My anxiety about carrying on a conversation with her disappeared. I was not used to such openness and felt a bit uncomfortable as she talked about her family. But I listened with great interest. She had three fathers. Her mother, Ruth, had married her high school sweetheart, Jim Perrin, but the marriage did not last long. Wes was born during Ruth's first marriage. After Jim, Ruth had been married to David Ray, an English professor, for a few years before that marriage too ended. David had adopted Wes. He was currently in Jaipur as a visiting professor at the university. David's wife, Judy, her daughter Sapphina, and Wes were going to be in India for the academic year. At that point, Ruth was married to the famous American poet Robert Bly. My Indian sensibilities shocked, indelicately I blurted out, "Why did your mother marry three times?" She did not seem to notice my disapproving tone.

"It did not work out. She was too young when she married Jim. David and she did not get along. She and Robert are happily married."

"What happens if it doesn't work out with Robert? Would there be a father number four?" I was about to ask but realized the question was inappropriate, one reflecting a hidebound Indian perspective. We talked for some time until she decided to leave. It had been a pleasant visit, and I was sorry to see it end. I offered to escort her back to her home. To my surprise and delight, she agreed. Wes and her family had moved to a rental house from the university guesthouse. A-15 Vijay Path in the adjacent locality of Tilak Nagar was exactly one kilometer from my home. She lived barely a ten-minute walk away; it took less than five minutes on a bicycle to get to her house.

We must have been an unusual sight. As we approached the slum at the end of Street No. 6, naked slum children, their bodies caked with dust, ran after us, laughing and shouting, "*Angrez, Angrez!*"* Foreign tourists were a common enough sight near Jaipur's forts and palaces, but it was rare to see one in Raja Park. It was even more uncommon to see an Angrez woman walking with an

---

* *Literally, this means an Englishman. Angrez is a catchall term for white foreigners in India.*

Indian man in suburban Jaipur. Many heads turned as we walked to her house. I was excited to be in the company of a lively, warm, and kindhearted girl who, incidentally, happened to be quite attractive. I knew that walking a young Indian girl who wasn't a relative of mine to her home would've been unthinkable. The thought made me hesitant to step inside her house when we got there. I quietly followed her inside, feeling uneasy. David, standing in the living room, welcomed me with a charming smile and shook my hand. Judy and Sapphina also happened to be home. It was a new experience: they talked to me as their equal and treated me as an adult. While Amar Singh treated me cordially and respectfully, I was always aware of my rank in his company. Feeling comfortable with the family, I invited both Wes and Sapphina to visit before leaving.

***

THE MEN AT THE STADIUM were not the only ones who thought that Western women were an easy conquest. The sophisticated, the educated, and the rich shared the same belief. Even if I was to try to befriend her, I was up against tough competition and impossible odds. Well-dressed, privileged, rich, and handsome young men who were fluent in English invariably gravitated toward her at the university. They flirted, they asked her out, they invited her for visits, and they offered her rides in their cars or on their motorcycles. She paid no attention to them. She probably did not intend to have a boyfriend in India.

Very shortly after I started my German classes, the professor showed us a German movie; all the students had to attend. As we stood around waiting for the technician to set up the screen and thread the 35-millimeter film through the projector, the students, especially the guys, were trying to engage Wes. For some reason, she hung around me; perhaps she thought I was harmless or simply felt comfortable with me; perhaps she could read that I had no agenda. I ended up sitting next to her. When we mounted our bicycles after the movie, it was dark. The roads were empty; there were few, if any streetlights. I offered to escort her home after the movie. We bicycled back together, taking a shortcut through an unlit pedestrian pathway between the sand dunes and the paved street.

A week later, out of the blue, Amritjit Singh, an English professor who was helping David settle down in Jaipur, invited me to a movie at the Mayur cinema hall, one of the two movie theaters in Jaipur that screened Hollywood movies on weekend mornings. Wes and Sapphina would be there, too, Amritjit told me. His offer put me in a difficult spot: I was perpetually short of cash, so where would I find the money for the movies? Rajesh came to the rescue, being more than happy to accompany me and pay for my ticket. His intentions were clear: he thought he had a chance to flirt with two American teenagers. His wishes did not come true. Wes and Sapphina sat to my left, Rajesh was stuck sitting with Amritjit. His fancy clothes, his clipped English, and his car did not impress the girls.

Going to this movie was a turning point for both Wes and me. I was a provincial unenlightened seventeen-year-old when I met Wes. My face was free of any hints of facial hair. The farthest I had ever traveled was to Delhi, six hours from Jaipur in a bus. Wes, two years older, had traveled the world, having spent part of her childhood in Europe. She was worldly-wise, I was unworldly; outgoing, extroverted, and gregarious, she laughed easily and loudly, I was quiet, introverted, too self-conscious, and inelegant. Yet somehow she liked to be with me. We began spending time together. In the beginning, Sapphina accompanied us everywhere. We bicycled to the old palace-fort of Moti Dungri near Wes's home; we strolled in the nearby parks, enjoying the landscape covered with rows and rows of beautiful flowers; we went to the forts and palaces of Jaipur. Both Sapphina and Wes regularly visited me at my home; I began to visit them.

Wes and I started running together at the stadium in the evenings. Then we started going to the stadium twice a day. In the cool and crisp air of the early morning, before the sun rose on the eastern horizon, I was on my way to her home. She would be waiting for my gentle knock at the door, and we were off to the stadium. After returning from the workout, I went to the college; as soon as I was back from the college, I was at Wes's house again, and we were off to the stadium for our second workout of the day. I lived in an old-fashioned home, in an old-fashioned neighborhood; this relationship was incredibly improbable, it was not supposed to happen or be tolerated! No unrelated male or female was allowed to meet openly in the orthodox society I lived in. We were such a contrasting pair; we stood out anywhere we went, ruffling feathers, evoking people's disapproval and ire.

On weekends, we went for excursions in Rajesh's car around the city. Despite Rajesh's firm belief that he was a ladies' man, his charm did not work on Sapphina or Wes. Rajesh was greatly offended that Wes and Sapphina focused on me during our outings, making him feel unwanted, destroying his confidence. The prince was chauffeuring around a pauper and the ladies; the ladies were ignoring the prince and paying attention to the pauper. Rajesh's techniques were not working.

One afternoon, after we returned from a Hindi movie, Rajesh looked sullen. After we dropped Wes off, he complained, "This is so insulting. Both of you ignore me as if I don't exist!" I couldn't help it if I focused on Wes; she was our guest! In India, guests are treated as God; furthermore, she was an attractive young woman. If our esteemed guest wasn't interested in carrying on a conversation with Rajesh, what could I do? Somehow she always ended up using me as a shield against Rajesh. Clearly she would rather have conversations with me than with him. He was frustrated and didn't want to try his luck anymore. I was relieved. I didn't know how to diplomatically tell Rajesh that neither Wes nor I welcomed his company. I did not want any man with designs to be around Wes. I was not going to allow anyone to take advantage of her.

———————

THE FIRST WEEK WHEN WES and I started training together at the stadium, I received an unusual message. Pratap, a fellow runner, stopped me as I was doing laps around the track. "Hey, Sunil, Krishna wants to see you right now." Krishna was the Asian Games gold-medal-winning long-distance runner. Krishna was the celebrity with a giant ego; always surrounded by sycophants, he paid little attention to the lesser folks. I hated celebrity worship and kept my distance.

Krishna was busy talking to two of his biggest fans—Om Prakash (OP) and TB. TB was the son of a powerful politician; his father was a legislator. I could never figure out if TB was born a thug or his politician father had turned him into one. In India, elected politicians and their families often act more like lawless criminals. OP was another thuggish athlete with political connections. The three stared at me as I approached.

"Did you want to talk to me?"

"Do you know your *aukat*?" Krishna said.

Asking about someone's aukat is a condescending way of indicating that a person is of a lower social rank and is acting too big for his britches; it was a challenge to fight.

"What?" I was taken aback.

"Who allowed you to run with that *gori* (white woman) here? Why are you licking her ass?" TB barked. He stood within inches, staring me down. "Do you care for your life? Do you want to keep your arms and legs intact?"

TB was a violent hooligan. With his political connections, he could literally get away with murder.

"I don't want to see you with this gori again," ordered Krishna.

"We are going to walk to her. You will introduce her to us and leave," TB said. "In future, you will stay away from her *unless* you don't love your life," he emphasized.

"Get walking, we will be there in a minute," OP chimed in.

I walked away in disbelief. In the middle of a stadium filled with hundreds of athletes, cricket players, boxers, gymnasts, hockey players, and weight lifters, I had been threatened and ordered to be the pimp for a white teenager, as if she were some disposable property I could dispense with. What were these bullies expecting? To take her to one of the rooms and rape her? Or demand that she become their concubine?

I was not going to introduce Wes to these degenerates or jeopardize her safety. I briskly walked to the grass patch where Wes was stretching.

"Let's go, Wes."

"I just started stretching."

"No, we need to leave right now."

"Why? I just started stretching."

"Let's go, Wes! We need to leave right now," I almost screamed at her.

She gave me an inquisitive look, "What's going on? What happened?"

Why doesn't she understand we are in trouble? I felt like screaming and dragging her out of the stadium.

"Please, there's an emergency. We need to go." From the other side of the track, Krishna and his hooligans were walking toward us.

"Let me use the restroom first."

"Wes, we have no time."

We jumped on our bicycles and sailed out of the stadium.

I was flustered and incensed. My mind was racing; I had to make some excuse to Wes and get us out of this tricky situation.

"Are you listening? Why did we have to leave suddenly?" Wes's question pulled me from my daze.

"Nothing too important, I have to go home to check on something." I tried to play down the situation.

"Then why were you so desperately trying to leave the stadium?"

I shrugged in response.

"I don't get it. You are strange," she said.

We stopped in front of her home. As she opened the gate, she turned toward me and asked if I wanted to come in. I decided to stay. Wes, Sapphina, and I hung around until it was dark; it was way past dinnertime when I arrived back home.

"There you are," my father VK said. "Three people were here to see you; they just left on their motorcycle."

Krishna, OP, and TB were pissed off that I had defied them; they had come to my home and waited for me to return. By chance I avoided them, but a confrontation was inevitable now. I didn't want to find out how far these bullies were willing to go. I didn't want to tell VK or Raju anything. My father would have asked me to stop seeing Wes after hearing of this incident; I had known a gori a few days and trouble had followed me home.

Later that evening, in desperation, I biked to my friend Rajvir's apartment downtown, which he shared with Mahindra. Rajvir, a marathon runner, was from Haryana; Mahindra, a 10K runner, was from Punjab. Both worked for the Indian Railways as athletes.

"Krishna and his friends are telling me to not be with Wes. They are threatening me."

"Don't worry at all, we will take care of it. You go home and relax. Keep your head cool, okay?" Mahindra said reassuringly.

"Yes, don't worry, we will talk to them," Rajvir repeated.

As I was getting ready to leave, Mahindra advised, "Don't go to the stadium tomorrow."

———————

DAVID AND JUDY WERE TRYING the indigenous experience in India, living a simple and Spartan life during their ten-month stay in Jaipur. Their home was minimally furnished, with no television, radio, car, or telephone—just books, lots of books. Their teenaged daughters, not used to living like renunciates, were bored. They wanted to do something fun. Fun was a good excuse to avoid going to the stadium the next evening. We went to the fair in Ram Niwas Bagh, the large park near downtown.

Like every public space in India, the fairground was packed with people. Blaring speakers, strobe lights, screaming children, shows, rides, creaking Ferris wheels, and trained monkeys doing tricks—everything was designed to attract the attention of fairgoers. The odor of spicy fried snacks permeated the air. We slowly waded through the crowd. Wes and Sapphina were excitedly looking around; everything was a novelty to them.

As we walked past the stalls selling trinkets, someone tapped at my arm. A squat middle-aged man, his thick goatee streaked with gray, dressed in a gaudy silk kurta and churidar trousers, his head covered in an embroidered white Muslim cap, was holding on to my arm. He looked repulsive.

Smiling lewdly, he said in Hindi, "Hey, brother, why don't you let me have a taste, too?"

I was left speechless by his brazenness. He shoved a wad of rupees in my hands. "Come on," he said accusingly, "why don't you share them? We should also be allowed some fun!"

In disgust, I jerked my hand out of his grip and walked away.

I understood that we were a rare sight—a tall skinny brown Indian teenager and two white female teenagers in tandem. We attracted attention wherever we went. At the fair, heads had been turning toward us nonstop. But I simply couldn't fathom how a stranger, in the middle of a multitude of people, had the audacity to stop me and demand that I sell my friends for sex. Had I been escorting two Indian teenagers, no one would have made such a proposition.

I was shaking in fury as I walked to catch up with the girls. An obviously rich, conceited young man stepped in front of me, blocking my way. He was young, handsome, and sharply dressed. He addressed me in English.

"I am Vinod. I am in the medical school," he introduced himself. "You are going to introduce me to these girls. If you don't, my friends and I are going to kick your ass. You won't get home in one piece." He pointed menacingly toward his companions.

I did not go to the stadium today, but the stadium had come to me.

Wes and Sapphina had stopped and were looking back at me. I heard someone shout my name. "Hey, Sunil, how are you?" It was my high school classmate Ashok.

Ashok, tall, dark, and muscular, was the son of a wealthy landlord. He looked completely wasted. As he staggered toward me, his gang of hangers-on was trying to keep him on his feet. Ashok the Arrogant, as everyone called him, had always treated me with respect.

"Hey, Ashok! What are you doing here?" I exclaimed. "Let me introduce you to my friends here." I pointed toward Wes and Sapphina. The blustering bully boy saw what he was up against and quietly retreated. Ashok shook hands with the girls and staggered away.

My evening had been ruined, my inner peace shattered. I felt humiliated and trampled upon. I had always enjoyed my friendship with Wes, but it had turned into a heavy burden; was she worth the aggravation and indignities I had to face? Before anyone else could accost me again, I told my goris that it was time to return home. I didn't tell them why, not wanting to ruin their evening. I was loath to poison their minds against all Indian men, marring their stay in India. Not all the men could be like that. Or could they?

It was not Wes's or Sapphina's fault that the perversions of Indian men saw them as sexual objects who could be coldly used, exploited, and discarded without a thought, without consequences. These goris had done nothing to deserve such treatment. They were not looking for boyfriends or sex in India; not promiscuous or easy, they were decent, thoughtful human beings who just wanted to be left alone. I had known Wes for less than a month. She was kindhearted, understanding, and considerate; in my opinion, her warmhearted temperament and naiveté made her vulnerable to exploitation. I admired her humaneness and was thankful for her companionship. I was not going to allow deceitful men to take advantage of her.

I had survived the fair. I don't know what transpired between my friends and Krishna and his lackeys, but no one threatened or hassled me for being Wes's training partner at the stadium anymore. But it did not stop incessant pestering. People I did not know warned me that she was going to destroy my life—after all, there's no creature more sinful than a woman. A gori could only be an energy-draining, immoral influence! I was cautioned that she was going to suck the life force out of me; almost every day I was asked if she had given me her pussy. Had I been chaperoning an Indian woman, no one would have the temerity to talk like this.

———————

A MONTH AFTER I MET this blond stranger at the stadium, we had become best friends, spending all our free time together. Wes was like a breath of fresh air in my life. Open, honest, and genial, she was not judgmental and did not look down upon the poor or the disadvantaged. And she was not put off by my lack of money and material possessions.

She told me about her serious relationship with Steve Booth, her boyfriend she had known since high school. She recounted stories of her family and confided in me about her trips to Europe with Sam, her younger brother, and her mother, Ruth; her desire to be a working actress; her conflicts with David. I mostly listened; I was content just to be in her company. I was too embarrassed to share stories with her about my dysfunctional joint family. Once she knew I did not get along with my relatives living in the same house, she was considerate enough to not probe any further.

We religiously went to the stadium twice a day. Going to the movies became our favorite pastime. Both Wes and I loved romantic Hindi movies. I translated the romantic dialogue and songs, although she could figure out what was going on without understanding Hindi; after all, nearly every Bollywood film had pretty much the same storyline! Wes somehow figured out that I never had any money. She graciously paid for me at the movies, without ever making me feel uncomfortable. It was awkward, but my desire to be with her overcame my uneasiness.

Wes and Judy began learning a classical Indian dance form called Kathak at the Jaipur Kathak Kendra, a dance school barely a mile from my college. In the afternoons, I started visiting, watching them slap their feet on the dance floor to the rhythmic beat of the tabla drums. So far I had seen her only in tracksuits or nondescript jeans and T-shirts, which managed to mask her beauty. Once she began to learn Kathak, she had some Indian salwar kameez dresses tailored. That's when I discovered how lovely Wes looked in Indian clothing. I never told her, though; it might have been presumptuous and chauvinistic. I also discovered something unnerving: I had begun to lose my appetite; pretty soon I could not eat at all.

I had been training hard twice a day, six times a week. Besides running, I had been training with the local boxing team. I needed mega amounts of calories, but however much I tried, I simply couldn't eat anymore. I was losing energy and weight. I told my boxing coach Mohan Lal, a flat-nosed middleweight pugilist. He gave me a hard long look and said, "Dutta, you better not be falling in love. There is no other reason for your melodrama."

"What are you saying, sir? I am *not* falling in love with anyone, sir!" I protested.

"I know what is going on in your mind," he retorted. "You will become crazy if you don't watch it."

What did my not being able to eat have to do with falling in love? With who? Wes? How could I fall in love with her? There were a million reasons *not* to fall in love. She was just visiting the country. I *was* protective and tried to shield her from having bad experiences in India, but what did that have to do with being in love? She was my buddy, my responsibility. Above all, she certainly wasn't in love with me; she had her own boyfriend to go back to. My inferiority complex prevented me from thinking of her in romantic terms even in my youthful flights of fantasy. She was the unattainable princess. Princesses don't fall in love with their servants; such stories are reserved for the Bollywood movies. I was a grown-up, too rational to fall in love; why would I become a Devdas voluntarily and spend my life weeping? Didn't love end in tragedy? The coach was being preposterous. I should have gone to a doctor for advice.

But the boxing coach turned out to be correct. I had gone crazy; I just did not happen to know it at the time. The implausible friendship had transformed into a fairy-tale romance. How did it happen? When did it happen? Neither of us knew. But it did happen, it was achingly beautiful, and it consumed our lives. The early-morning and evening rides to the stadium, the afternoon visits to the Kathak Kendra, the evening visits to each other's homes, the uninterrupted togetherness on the weekends—Wes and I were already spending all our free time together. But it was not enough.

We were not content; our longing for each other was a deep constant ache. Even when we were together, we worried about the evening, when the cruel night would bring separation. We had to create more hours in a day. I started ditching my classes; then without telling anyone, I stopped going to college altogether. Wes let go her German class and her dance training. Now we could be together the whole day from morning till night, but that was still not enough. We stopped going to the stadium. It was hard to hold hands or stare at each other while running or training.

When one is in love, the outside world ceases to exist. Wes and I became defiant of our families and the society we were living in. We had eyes only for each other. We jettisoned two bicycles and rode together on one. Though she exuded affection and love and kept telling me every few moments how much she cared for me, I could not tolerate her speaking to anyone else or giving any attention to others. She relished my possessiveness, though sometimes it became a source of our lovers' quarrels. Unconcerned with the orthodoxies of society, we scandalized others by holding hands. We asked strangers to take our pictures as we held each other affectionately. We were inseparable, lost in each other. Alas, that time of blissfulness was destined to end. Suddenly it was May 1982, the date of Wes's departure back to the United States but four weeks away. As the dreadful day approached, we clung to each other physically and emotionally, even more tightly than before, as if that were possible. The fairy tale of unfettered love was coming to an end; love was going to end in separation. The test of true love was about to begin.

6

FALLING OUT

Don't break off the ties we have,
If nothing else, let's be enemies.
— GHALIB

In 1981, Raju had turned twenty-one. He and Amar Singh, who was then fifty-six, had become close friends. Amar Singh had grown very fond of Raju; they went to movies, shopped together, ate together, and sat and talked for hours. Amar Singh bought gifts for Raju. I noticed this with envy. Though Amar Singh was always friendly and treated me with respect and kindness, in the ten years I had known him, he had not *once* offered me a cup of tea. If I visited his house and Amar Singh was eating, he did not invite me to join him. To offer food to visitors is an intrinsic part of Indian tradition. Amar Singh *half* followed the tradition: he offered food to Raju.

Sometime in late 1981 Amar Singh started calling Raju his son. He had notices distributed on the MI Road, the fashionable boulevard in Jaipur, and in Raja Park, announcing that he had adopted Raju. He also declared that he was going to get his adopted son married. Raju was now writing his name as Kaushal Singh Dutta. Amar Singh also employed Raju as his secretary.

My parents were greatly relieved. They were concerned that Raju would never amount to anything, becoming a burden in their old age. Raju had shown little aptitude for studies and was interested in mischief instead. If Amar Singh was taking care of Raju as a son, what could be better than that! If Amar Singh wanted Raju to get married and pay for the wedding, my father VK was more than willing to help. VK began a diligent search for a suitable girl for Raju.

Raju did not have to suffer the deprivations of his family anymore; he could eat better, dress better, and enjoy his life better. He could also feel proud that he was associated with gentry, in contrast to his natal family of crude and unrefined peasants. And Raju was happy. He had new clothes, a new motorcycle, money to spend, and above all, a shotgun! Guns, a big status symbol, were rarer than cars in India.

———————

IN THE SPRING OF 1982, Amar Singh was upset about the death of Rajesh's cousin, Manvendra Singh, in New Delhi. Manvendra, a tall, athletic, and handsome teenager, had crashed his motorcycle into a bus; he was killed instantly. Rajesh had bought the motorbike Manvendra had crashed and had given it to Manvendra, against the wishes of his family, so everyone blamed Rajesh for his death. Amar Singh was getting an earful from Manvendra's family. Amar Singh, who did not think highly of Rajesh even before Manvendra's death and often complained that Rajesh was lazy and soft-witted, was now enraged at Rajesh.

But the tragedy of Manvendra was not the only cause of conflict between the father and the son. Raju was an even greater sore point. Rajesh hated the fact that his father spent money on Raju and treated Raju better than him. One afternoon he exploded in front of me, loudly cursing his father. I couldn't believe what I was hearing.

Furious, spittle flying from his mouth, Rajesh denounced Amar Singh, calling him a debauched homosexual and a disgrace to the Rathore clan.

"He is a shameless degenerate. That's why no one likes him in the house of Bikaner; he has made us outcasts. He has squandered our family wealth and treasure. Instead of living in a palace, look where we live! He has thrown away more than 1 crore [10 million rupees] to finance his debauchery," Rajesh fumed. I was taken aback at what Rajesh said next. "Raju is taking advantage of him."

Rajesh stared at me accusingly, as if somehow I was to blame for his family disputes. I stood up and quietly walked out.

More than a thousand years before the advent of the Internet's biggest attraction—pornography—Hindu temples were decorated with erotic sculptures and drawings on their walls that could teach porn actors of today a few things. The treatise on the art of love, lust, and sexual positions, the Kama Sutra, was completed approximately two thousand years ago on the Indian subcontinent. Yet while graphic illustrations of sex in centuries-old Hindu temples in India are part of Indian civilization, discussion of sexuality is taboo in India.

If sex remains a forbidden and shameful topic in India, homosexuality is detested and reviled. Despite a glacially slow change of attitudes toward homosexuality in large metropolitan areas, Indians remain intolerantly homophobic. Gay people are subjected to ridicule, humiliation, and violence. Only Sufi saints or the powerful in the Indian subcontinent can escape condemnation for being gay.

The Duttas were neither powerful nor eminent Sufis. Being associated with a homosexual relative would have caused scandal and condemnation. However, this was not my concern. I was more upset because I saw Rajesh's outburst as the ending of our friendship.

I did not know if Rajesh was right or just lashing out. Nevertheless, if Amar Singh and Raju were having an affair, how could that be my fault? Raja Sahib was fifty-seven, thirty-five years older than Raju; how could a provincial boy take advantage of a savvy man?

———————

AFTER WALKING OUT OF RAJESH'S room, I remembered an incident that had happened a few years ago. It was a quiet and pleasant Jaipur evening, though I was not enjoying myself. I had finished the books and comics I had borrowed from the library. The radio was playing Indian classical music, sounds that gave me the same joy as the bleats of a goat. I decided to visit Raja Sahib and see if he was in the mood to tell me an interesting story.

A driverless rickshaw stood in front of Amar Singh's house. Perhaps someone was visiting and the driver was helping with the bags. I walked around to Amar Singh's room toward the rear of the house. He was not there. I knocked and went in through the rear door, which led to the dining area and the living room. Amar Singh sat at the dining table. Across from him sat a sweaty man dressed in ragged clothes. The gaunt and exhausted-looking rickshaw driver glanced at me with a blank expression and turned his head away. He was eating while Amar Singh watched.

In the stratified, hierarchical, rank-, caste-, and class-obsessed Indian society, no one, not even a poverty-stricken refugee, would have a rickshaw driver step inside his home, much less have him sit at the table. Despite our penurious existence, I was thoroughly conditioned to feel proud of the fact that I was a Brahmin; we may not have had much, but we were better than others by birth alone. I was shocked; Raja Sahib was miles above me in social rank and privilege, yet he was feeding a rickshaw puller in his dining room! I was so at a loss for words, I forgot to greet Amar Singh.

Amar Singh, who had been scrutinizing the rickshaw puller

closely, turned toward me. "Why don't you leave now? *Mhare des se koi aya hai.*" A visitor has come from *mhare des*—meaning my native land, my hometown. *Mhare des* was a convenient explanation, since it required no elaboration. It was sufficient to claim that someone was visiting from one's native land without revealing a relationship. I nodded and headed out the back door without saying anything. I opened the gate and slowly walked out, staring at the rickshaw. In shock and confused, I aimlessly strolled on our street, feeling as if I had been wronged.

I identified Amar Singh with royalty and took pride in knowing him. He was a wise man, my role model, someone who towered above the ordinary mortals in my opinion. My role model had turned out to be a human being like the rest of us plebeians. Ghalib had once warned:

> For the love of God, please don't lift the curtain over the
> Kaaba.
> Perhaps in that spot we may find an ordinary stone.

The veil had been lifted. The prospect of Amar Singh's being gay was unsettling. I had heard that homosexuals were sick people. They were abnormal, perverted, and immoral—a danger to society. The highly offensive colloquial term for gays in India, *gaandu*, is derived from the Hindi word *gaand*, meaning ass; thus it signifies someone who practices anal sex. Was Amar Singh a gaandu? I had known him for some time now; I knew he was not abnormal, perverted, or dangerous in any way! Though I was envious that he lavished all his attention on Raju, he was a gentleman who had always treated me with kindness.

Later that night, I sneaked out to spy on Amar Singh. The rickshaw was not parked on the roadside anymore; it had been moved into the driveway of Amar Singh's house. Early in the morning, the rickshaw was still in the driveway. When I left for school at seven

A.M., it was gone. This was one of the many times I saw Amar Singh bring socially and economically disadvantaged men to his house for overnight stays. I quietly accepted the fact that Amar Singh was gay and gave up my stereotyped notions of gays. What others in the world I lived in believed about homosexuality and homosexuals was wrong; it did not accord with the Raja Sahib I knew.

———————

IN JULY 1982, I LEFT for Hisar, a town in the state of Haryana. I was finally admitted to Haryana Agricultural University (HAU) after the disappointing bureaucratic snafu of 1981 which kept me from joining HAU. It had not been a loss in any way; instead of school, I had found love! VK came to help with my move, which consisted of carrying one bedroll and a suitcase, the entirety of my possessions.

A month after I left for Hisar, Amar Singh sold his house and bought two adjacent houses, 2/63 and 2/64, a few kilometers away in Malviya Nagar, Jaipur; he gave one to Raju. Amar Singh's benevolence was not limited to giving Raju a fully furnished house. On August 18, 1982, he opened two accounts at the State Bank of Bikaner and Jaipur, Gandhi Nagar branch, Account No. 2182 for himself and 2183 for Raju. That day he transferred 100,000 rupees to Raju's account from his account. Amar Singh also purchased expensive jewelry for Raju. After the move, Raju seldom visited our parents.

Amar Singh traveled with Raju to Bikaner; they stayed at the Lalgarh Palace, constructed by Amar Singh's grandfather, Maharaja Ganga Singh, now a heritage hotel. This was the palace where the viceroys and governors-general of India used to visit and stay during the colonial times. Amar Singh was born and raised in Lalgarh Palace. Amar Singh used to own a part of the palace. Raju met many from the Bikaner's royal family.

At the time, I wasn't aware that Raju had left Raja Park with

Amar Singh. I never heard from Raju or my parents about his move to Malviya Nagar and his new life. Raju had always been standoff-ish with me. After he moved to his new house, he did not invite his family for a house-blessing ceremony. If our parents had any hope of relying on Raju in their old age, their hope must have been dashed.

But life moves in mysterious ways. In October 1983, just one year after he had left with Amar Singh, the adopted son was back in Raja Park. He did not have his house, his motorcycle, his jewelry, or his fat bank account. I learned about it when I visited Jaipur during my winter vacation that year. Something disastrous had oc-curred, and neither Raju nor my parents would discuss it with me. I decided to seek answers on my own.

I went to see Rajesh in his apartment in C-Scheme. We were not the friends we used to be, but our relationship was not over. Rajesh was guarded but not unfriendly.

"What happened between your father and Raju?" I asked.

"He *really* upset my father," Rajesh responded. "It is not easy to make him angry, but Raju did!"

"What could have happened? Rathore uncle was the one who gave Raju a house and a motorcycle!"

"Raju was too obsessed about knives and guns. He always talked about them. My father just couldn't handle it anymore," said Rajesh.

Though our years of association, I knew that Rajesh was prone to exaggerate and misrepresent to make his point. Was it possible to upset Amar Singh by talking about guns and knives? Raju loved guns, but Amar Singh had purchased a shotgun for Raju in the first place. I had seen two of Amar Singh's biological sons fight with their father and leave the Raja Park house. Now the adopted son had been stripped of everything and booted out. As I left Rajesh's place, I still had no real answer as to why Amar Singh and Raju had had a falling-out.

IN HISAR, I WAS AN outsider. The dominant community, the Jats, were a proud people, certain of their special place in the social hierarchy. The Jats condescendingly called Punjabis like me *riphoo-gee* (refugees), a term reflecting the migration and settling of the Punjabis and the Sikhs in Haryana after the 1947 partition. The prevailing stereotype was that riphoogees were lazy, vulgar, un-wanted, uncultured, uncivilized, cutthroat, and selfish—all in all, a danger to Haryana's culture. They were the maligned aliens within their own country. I had moved from a refugee colony to a univer-sity campus; my status had not changed much. Fortunately, I was spared the put-downs and disdaining looks; since I came from Jai-pur, people presumed I was not a Punjabi. I felt no need to correct their mistake.

I was surrounded by students from conservative rural back-grounds. I looked different, dressed differently, came from a dif-ferent background, held different values, and spoke a different language. I also let my college mates hear my idealistic contrar-ian views on the immorality of the caste system, the subservience of women in Haryana and Rajasthan, and other topics. People laughed, shook their heads, and ignored my arguments: I was a city boy, so what did I know about tradition? Unbeknownst to me, I had become infamous for a very different reason.

I had been waiting to hear from Wes since her departure in June. Despairing, forlorn, dejected, every few days I sent her a let-ter at her mother's address. I didn't know if she was back in college in New York City or elsewhere; I knew nothing! What happened to all the promises, the tender embraces, her unceasing declara-tions of love? Hope was lost; nothing remained except pain and memories.

As a freshman, I shared a men's dorm with two classmates. I refused to sleep on a bed; I slept on the floor and divested myself of most of my belongings. I was trying to live like a monk. Besides my clothing and college books, all I had was a stack of photographs of Wes and me and an envelope containing a lock of her hair she had given me before leaving. My classmates ornamented their rooms with glamorous alluring starlets; I pasted a blowup of a photo of Wes on the wall above my tiny desk. Perhaps my eyes gave me away, perhaps my lovelorn pining was too obvious, but my classmates figured out that the picture was an object of adoration and devotion, not an ornament. As an eighteen-year-old separated from his American girlfriend and in an unlikely long-distance relationship, I was an enigma to them.

The school year had just begun, and I was slowly getting to know people. One day I was visiting my new friend Jagdeep. Jagdeep was a popular and well-connected veterinary student. HM, the president of the HAU student union and a slick young Jat politician, and several of his acolytes showed up at Jagdeep's room for a visit at the same time.

"Have you heard about this Majnun* (crazy in love) in the college of agriculture? He sleeps on the floor and cries for hours in front of his Layla's photo. He has pledged not to sleep in a bed until he sees his Layla again!" HM said to Jagdeep, feigning astonishment.

Jagdeep smiled, pointed toward me, and said, "Here is that Majnun."

The others burst out in loud cries of "Oh, *he's* the Majnun." I blushed in embarrassment, angry at Jagdeep for exposing me. The only truth to the rumors was that I did sleep on the floor. I had been in Hisar for barely a month and everyone in the campus was talking about the freshman who had gone crazy in love.

Every day after lunch at the cafeteria, before returning to the

---

* *Layla and Majnun is an ancient legendary love story, popular in South Asia and the Middle East.*

classes, I walked to my room in anticipation. As I reached my room, I would slow down, with my heart pounding against my chest wall. Was this going to be the day? Was I going to find a letter from Wes? Every day my hopes were dashed. Every day I stood quietly inside the room—angry, disappointed, glum, unsure. Sometimes I went to the campus post office to check my mail, as if walking to the post office would work out a miracle. The result was the same. I felt helpless, forsaken.

Whom could I blame? A faithless, fickle young American girl? Was love so trivial for her? She left, and her promises disappeared? What was I to make of her ignorance and lack of concern about the man she used to shower every day with declarations of love and hold tightly? For months we spent every day together, from morning till late evening. I had given up boxing and running to be with her all the time. She had given up her German class and Kathak dance training to be with me the whole day. We had been inseparable, yet now we were separate!

As the Punjabi Sufi saints have said, true lovers spend every day prostrating themselves again and again in the footprints of love and memories. Does a man in love abandon fidelity and love because of adversity or rejection? No, the lover within me was not going to give up at any cost. *Whether you listen or not, I will tell you the condition of my heart.* I tried to express my feelings in the long, loving, yearning, angry, sad, reproachful letters I sent her every few days. While my words alone could not convey the state of my heart, great Urdu poets lent a helping hand with their spiritual love poems.

———

I NEEDED TO GET MY mind off Wes. I started running and training with the university boxing team twice every day. Early in the morning, we ran, did calisthenics, and shadowboxed; in the evening, we sparred. At the end of the day I was wiped out. Exhaustion is a great antidote to emotional pain; like an opiate, it distracts, it numbs, it masks; like an opiate, it has no effect on the underlying affliction.

Every time I checked my mail, I was prepared to be disappointed, and my beloved never failed to disappoint. This became a daily ritual. I was losing hope of ever hearing from her again, yet I couldn't let go. One must remain wary of hope; it leads only to regrets and disappointments. One afternoon three months after Wes had left India, I walked to my room after lunch. I opened the door and looked down, and my heart trembled. There on the floor was an aerogram.

It was from America, but it was not from Wes.

The letter was from Ruth, Wes's mother. After weeks of waiting to hear from Wes, I had written an imprudent letter to Ruth, asking if her daughter was still alive. That brought a quick response; I had received a reply within two weeks. Disregarding my rudeness, Ruth had written back, thanking me for all I had done for Wes. She said that Wes's experience in India had been positive and loving because of me. Wes had been busy, changing colleges and moving to another state. I should hear from her soon. I *should*. Her mother had had the time to reply to my first letter as soon as she had received it! Wes had ignored a dozen . . .

Ruth's letter had given me hope. And the hope finally came to fruition. A few days later, I found a thick envelope, addressed in the familiar childlike printed handwriting—Wes could never write in cursive—waiting on the floor. One sight of the letter, and all the anguish, resentment, anger, and grief over abandonment was instantly assuaged! I read the dozen pages she had written, again and again. *I love you*, she had written; *I miss you*, she said again and again; *I am sorry for the delay in getting back to you*, she apologized; *I wish we were together*, she said. Sheets of paper covered with her writing compensated for her missing voice. I don't know how long I stood at the threshold of the door, holding the letter, reading, dreaming, remembering, in a daze. A piece of mail had revived my hope, and with that, the possibility of future disappointments.

Wes was slow. Simple daily tasks of bathing, brushing teeth, and getting ready took her hours. Maybe writing letters was even more difficult for her; maybe she was testing my resolve and patience.

———————

IT WAS TOWARD THE END of May 1984. I was in the botanical garden doing fieldwork when Ashwini called out loud, "Dutta, your brother is looking for you."

"Are you sure it's my brother?" I asked. Why would Raju visit me in this dull town? He was an excitement-loving city person.

"I told his driver to go and wait at your dormitory," Ashwini said. "I have been looking for you for an hour."

As we walked back toward the dorm, Ashwini exclaimed, "Your brother has a real fancy car." A sleek and shiny Japanese import, with a smartly dressed chauffeur, was parked in the portico. An idle bunch of students stood near the car, gawking. Raju was in the back seat; he had a big grin on his face.

He shook my hand, saying, "I don't have much time. I need to leave soon." I carried his suitcase to my room. His luggage was elegant, expensive, and imported. Raju was fascinated with foreign brand names, but I couldn't blame him for that. India was a protectionist nation, and the importation of consumer goods was highly restricted and heavily taxed. Indian factories, shielded from any potential competition, churned out poor-quality consumer goods. Those who could afford them sought superior-quality imports despite their high cost.

"What is going on? Where did you get this Japanese car? Is it yours?" I exclaimed.

"I just returned from Europe. I rented the car from the five-star hotel in Delhi I am staying at," Raju said as he opened his suitcase and started showing off what he had brought from abroad. The suitcase was filled with clothes, knickknacks, and fancy chocolates. "I have been in France, Germany, and Switzerland for the last one month," Raju told me. "Look at this!" He displayed a compact blue steel pistol and a box containing shiny bullets. I was unnerved. I had never seen a handgun. "Is it going to go off?"

"It only fires blanks."

I did not know the difference between real bullets or blanks, so what Raju was holding looked like a real gun and real ammunition.

"The customs at Delhi airport took away the real pistol I bought in Germany. I need a license to get it back," Raju told me.

Firearms, especially handguns, are strictly regulated in India. People couldn't possess a firearm without a license. It was far easier to attain God than a firearm license in India.

"All this must have cost a fortune." I tried to imagine how much Raju had spent. It was expensive enough for me to take a bus from Hisar to Jaipur; Raju had just flown back from Europe and traveled to Hisar in a chauffeured car! He was living the good life and enjoying it. I wondered how much money he had made in Amar Singh's employ.

"I am leaving for Europe for a job. I have some offers there. Keep this suitcase with you and I will get it back later."

Raju had always been fascinated with three things: owning guns, flying, and above all, migrating abroad. If there was a job waiting for him in Europe, it would fulfill his dreams. Raju headed back to Delhi in the evening, leaving his empty suitcase.

———

A FEW DAYS AFTER RAJU LEFT, I went to visit my parents. After the overnight trip, I took a rickshaw from Jaipur's main bus stand. As the rickshaw approached Raja Park, I noticed something was seriously wrong. The neighbors, instead of smiling and responding to my greetings, were averting their faces from me. When I reached home, my parents looked funereal. Anxiety and tension hung in the air. I had never seen VK forlorn before.

A few weeks ago, Raju had told our parents that he was going on a trip to visit his maternal aunts in Delhi. Four days later, a group of bank officials had visited my parents. Raju had forged Amar Singh's signatures and stolen 75,000 rupees from Amar Singh's account. The manager offered a deal to my parents: if Raju would return the money, the bank would not report the crime to the police. But Raju had disappeared with the money.

On April 14, 1984, Raju had audaciously walked up to the bank manager, identifying himself as Amar Singh. He asked to withdraw 150,000 rupees from Amar Singh's term deposit account. Unbeknownst to Raju, Amar Singh had taken a loan of 75,000 rupees against the term deposits, leaving only 75,000 rupees available. Undeterred, Raju asked to cash the remaining 75,000. The bank manager had a temporary memory lapse: how could Amar Singh forget he had taken a large loan against his term deposit recently? The manager summoned his assistant, converted the term deposit to a current account, and issued a checkbook to Raju on the spot. Sitting in the manager's office, Raju forged Amar Singh's name on a check for 75,000 rupees, pocketed the cash he received, and walked out.

Amar Singh and Raju were well known to the bank employees. The two went to the bank together, and Raju often received money on Amar Singh's behalf. Sometimes Raju even signed the check for Amar Singh. Bank employees knew him as the adopted son of Raja Sahib. The manager was new to the branch and didn't personally know either of them.

A few days after the visit of the bank personnel, the police showed up at our home. Amar Singh had discovered that his account had been robbed and reported it to the police.

The police in India are well known for illegal detentions, refusal to take crime reports, routine harassment of the accused and their families, coerced false confessions, intimidation, false arrests, false imprisonments, falsification of evidence, extortion, kidnapping, torture, and in some cases, murder. Extortion by the police, called *hafta* (weekly contribution), is not limited to condoning illegal activities; the police also extort those engaged in legitimate business: traders, truck drivers, rickshaw pullers, shopkeepers . . . Being honest and innocent does not protect one from the police. The rich and the elite receive special treatment and get away with rape and murder.

Raju had visited me six weeks after committing a felony; he did not tell me that he was on the run from the law. Ashamed of what Raju had done, my parents had kept me in the dark, hiding from me the fact that the police were harassing them.

My visit to Jaipur was a dismal shock. Raju had run away after committing a crime, and we were left to face the consequences. It was almost worse to face the scorn of society every day than to be in an actual prison. Every time I tried to make eye contact with our neighbors and friends, people turned their heads away. If you find yourself down, the "spiritual" Indian society would give you a kick in the ribs. People I had known all my life were bad-mouthing and avoiding us.

I could not remain in Jaipur long; I had to leave for the exams at the end of the quarter. I took the evening train from Jaipur, changing trains at two A.M. There were no direct trains to Hisar. It was about six in the morning when I reached my dorm. I had spent the night sitting on a wooden plank bench in an overcrowded compartment, dense with stale air. Sweaty villagers jammed against one another, occupying every possible inch of space with their bundles and baskets. Sand swirled into the windows as the train moved through the desiccated land of Rajasthan. Ash and embers from the steam engine found their way in through crevices. I was covered in soot and exhausted, ready to reach my room and collapse in my bed.

It was not meant to be.

To my surprise, the room was unlocked and the lights were on. Raju was waiting inside!

Dumbfounded, I confronted him. "What are you doing here? What happened to the job offer you had in Europe? And what did you do in Jaipur? You are going to get me arrested. The police are looking for you. They are harassing our parents. No one talks to us anymore because of what you have done. What the hell were you thinking? What is going to happen if the police are watching my room? Both of us will go to jail!"

I was racked with anxiety. He was nonchalant and unruffled.

"I was angry. I wanted to take revenge. Amar Singh owed me 150,000 rupees when I left his service. He didn't pay me. So I took the money from his account." Raju did not elaborate any further. He seemed convinced of his righteousness.

"Do you know the police are making life difficult for us? Didn't you know what was going to happen? We have become pariahs; Avinash and others are gloating. It is hell to be in Raja Park now."

"I was very angry. I had only one idea in my mind, to quit India and to look for a job abroad, at any price. I never thought of what will happen to the family and parents." Raju said matter-of-factly. At least he was straightforward.

———————

RAJU TOLD ME THAT WHEN Amar Singh hired Raju as his personal secretary and also adopted Raju as his son, Raju was promised a furnished house, a form of transportation, gold jewelry and gar-ments for his future wife, and a firearm. In case Raju ever decided to quit, he would get to keep the gifts. Amar Singh traveled with Raju to Bikaner and introduced him to relatives as his adopted son. Things changed in late 1983. Amar Singh told Raju that he was a homosexual and that he wanted to have relations with Raju. He wanted to maintain the facade of a father-son relationship for the outside world while having a relationship with Raju. Raju said that he was shocked and left Amar Singh the next day. A week af-ter Raju returned to Raja Park, he received a long letter from Amar Singh. The letter was full of romantic poetry, with a dejected Amar Singh begging Raju to return. Similar letters continued arriving every week until March 1984, when Raju went to Chandrashekhar and complained about Amar Singh.

Raju vehemently denied having an affair with Amar Singh. Raju claimed he voluntarily returned the house, the jewelry, the cash, and the motorcycle, and returned to Raja Park. When Raju asked for the 150,000 rupees promised in case of job termination, Amar Singh refused. In revenge, Raju cleaned out Amar Singh's bank account.

Amar Singh provided a statement to the police: "I left Raja Park in August 1982 after living there for 12–13 years. I shifted to Malviya Nagar 2/63 2/64, and this boy came with me to Malviya Nagar. Due to love and sympathy, and to help get him married, I kept him with me. When his parents did not find a girl for him, this boy returned to his parents' house in April 1983.* I have not seen him since." In his second statement, Amar Singh added that in April 1983, due to some misunderstanding, Raju became angry with him and returned to his parents' home.

This conflict between Amar Singh and Raju was mind-boggling.

---

* *Raju returned in October of 1983.*

I didn't know who to believe. I did not know what promises were made or broken; I didn't know what relations existed between Amar Singh and Raju. I did know that it was wrong for Raju to avenge any perceived betrayal or broken promises. Now his revenge had backfired and he was a fugitive. In addition, he had made our lives miserable and had put me at risk of getting arrested.

Usually the police were lax in their investigations; however, Raju had committed a crime against a powerful man, so the police had to show that they were diligently investigating. I remained on edge, worried that the police might show up at any time. Despite my concerns, Raju was not leaving. He had spent the loot on his Europe trip. I did not know what options he had but to give himself up to the police. He had something else in mind.

Raju was comfortable in the knowledge that I wouldn't call the police to report him—it would have been unthinkable to call the police on one's own brother. I didn't want him hiding in my room, but he was unwilling to leave or surrender to the police; we were at an impasse.

———————

RAJU STAYED IN HISAR FOR several weeks that summer. My friends were unconcerned that he was a criminal on the run; they were accommodating and sympathetic. In Haryana, a highly conservative, predominantly rural society, people backed their family members at any cost. The aspect of revenge in Raju's story had enhanced his status. Raju's outgoing nature and his remarkable ability to tell tall tales of adventure, intrigue, and beautiful women in Europe had made him popular.

I lived under constant stress that summer. After taking my final quarter exams, I left for Jaipur for the summer break; Raju stayed in my room. Some time during the break, Raju left Hisar to hide in a nearby village. When I returned to the college after the break, to my relief, Raju had disappeared. The relief did not last too long. Late one night, he reappeared.

He looked gaunt, as if he was not getting enough to eat. His cheeks were sunken and his shirt and pants hung loosely over his frame. His shiny dark hair had turned into a brittle and listless straw-colored mop. He was using chlorine to bleach his hair. I felt sorry for him. Surprisingly, he was in good spirits.

"I should be able to go abroad now," he said.

"What do you mean? They will arrest you the moment you show your passport at the customs!"

"That is not going to happen. Look at what I have."

Raju handed me a passport. I flipped it open. A blond man's picture stared at me. It was a Swiss passport.

"Did you steal it from some poor tourist? Have you done something wrong again? What are you going to do with this passport?" I felt sick.

Instead of answering me, Raju handed me a small metal embossed seal. I shook my head in resignation. How far was he going to go to escape from India? First he robbed Amar Singh and used the money to flee abroad. Now he had robbed some poor traveler of his passport and belongings. What made him think that another trip abroad would end up differently?

While staying at a YMCA youth hostel in Delhi, Raju had stolen a Swiss passport. He found a metal fabricator in Rampur to make a forged Swiss seal to alter the passport. He also had a rubber stamp made in Punjab to forge the passport. Now I had to deal with the growing list of his new crimes; besides robbing some hapless tourist and working with other criminals to forge the passport, who knows what else he could had done?

Could Raju's new escapades make our lives worse? Could we be shamed twice as much? Could our neighbors and friends, who had stopped talking to us and paying visits, shun us any more? Could the police threaten or hassle us twice as much, or demand double the extortion money for letting us alone? Things couldn't get any worse, I concluded.

IN THE 1980S, PASSPORTS WERE not as secure as they are now. In this case, the passport owner's photograph had a raised impression of a Swiss emblem embossed from behind the page it was affixed to, using a male and female die. It was also stamped. However, the photograph had no lamination or veneer over it. That summer, Raju used a sharp razor to painstakingly remove the original photograph layer by layer. Eventually the remains of the picture were peeled off; he affixed his own on top of the original. He then used the metal seal behind the page the photograph was glued to, to create the raised emblem on his photograph. This was tricky. Letters on the die had to fall in the exact spot where the original photograph had been embossed; any misalignment would have disfigured and damaged the page. After embossing the photograph, he stamped his photograph, superimposing the lines of the original with the fake stamp. The forged passport was ready.

I did not know if the forged passport was good enough to get Raju through the immigration and customs counter in Delhi airport. His bleached hair did not look anything like blond, he didn't have blue eyes, and he certainly didn't speak Swiss German or Swiss French. Maybe he will escape, maybe he will get caught. In either case, I will not have to live in fear anymore.

I hoped that Raju would escape and make a better life abroad. I wished that he would use his skills for better purposes in future.

ABOUT SIX MONTHS AFTER RAJU disappeared from Hisar, I was visiting Jaipur. One morning I waited at the front gate of my parents' house for the postman. Waiting for the postman had become my favorite pastime. He was my lifeline, bringing love letters from across the oceans. He knew it and used the knowledge to his benefit. A letter from America was not handed over to me without a baksheesh, a tip.

I saw the familiar bicycle approach. The postman dismounted, pulling an envelope from his tan canvas mailbag. He half extended his arm, holding an aerogram. Looking expectantly, he declared, "This is *foreign walla*." The letter was not from Wes and therefore didn't hold much appeal to me. Taking the envelope from his hand, I offered him half a smile. Letters were my sole means of communication with Wes. Calling abroad was too expensive and time-consuming, and often the calls failed to go through.

If I wanted to call Wes, I had to navigate through four miles of traffic to the downtown general post office. The trip was difficult. The roads were packed with pedestrians, bicycles, auto rickshaws, smoke-belching buses, tempos, camel carts, wandering cows, trucks, bullock carts, stray dogs, cart pushers, all moving in random directions; it was a challenge to steer my rickety old bike through the bedlam. My ordeal was not over once I arrived at the stale smelly soulless general post office. After waiting my turn in a long line, I would fill out a detailed form listing her name and phone number, include the purpose and expected duration of the call, hand the money to the clerk, and wait for the call to go through. The clerk placed the overseas trunk call (international call), which, for unfathomable reasons, took hours to go through—three hours seemed to be the minimum requisite time. I had to remain alert and on edge the entire time. My name could be called out at any point; missing it resulted in cancellation of the call.

In the prehistoric days before cell phones, Skype, and email, a phone call was an anxiety-ridden wait with no guarantee that it would go through, and if it did, there was always the question of whether Wes would be at home. There was a time difference of ten and a half hours between Jaipur and New York City; between school and work, Wes was hardly home. If I was lucky and all the heavenly bodies aligned properly, I would hear the clerk shouting my name, run to the glass booth, and pick up the phone. It was a mixed blessing! Call quality was routinely horrible, the sound muffled, and when the call did not terminate mysteriously, the clerk kept butting in. You can imagine how romantic it was when a stranger, monitoring your conversation, interrupted you repeatedly, asking how much longer you were going to be on the phone or whether he should terminate the call because it had lasted over three minutes. He also kept reminding you that other people were waiting in line, too.

Yet, it was blissful to hear Wes; her voice was the tender touch creating a bridge between my fond memories and the vast distance separating us. I wished we had a telephone at home and I could talk to her whenever I wished. Ownership of a telephone seemed to be the minimum essential requirement for a long-distance romance. Lacking that, I had to rely on letters.

THE AIRMAIL ENVELOPE THE POSTMAN had handed to me came from Canada. The content was puzzling. I did not know the sender, K. S. Sidhu, though his name was familiar. The return address on the envelope was from a detention facility in Vancouver. The message was cryptic: *My name is Kulwant Singh Sidhu. On my way to Canada, I met your brother in Tokyo. He is doing well; you don't need to worry for him.*

Who was this K. S. Sidhu? Where was Raju? Was Mr. Sidhu writing from a prison? How was he connected to Raju? Was Raju in Canada? Was he also locked up in a prison in Canada and too embarrassed to write himself? It took me more than a year to find out who Sidhu was and where Raju had ended up.

7

THE DANGEROUS GAME: RAJU WITH KHALISTANIS

By 1984, Punjab was in the grip of terror. Using hit squads and terror, Jarnail Singh Bhindranwale, a fiery fanatical provincial Sikh preacher, was running a parallel government from the premises of the Golden Temple, the Sikh religion's holiest shrine. Separatist Sikh militants were demanding Khalistan, a state for Sikhs carved from Indian Punjab. Rifle-toting Sikh terrorists were robbing banks, killing policemen, executing newspaper editors who dared criticize them, and murdering government officials. Trains were bombed, bullets sprayed in bazaars, buses were stopped; Hindus were marched out and executed in cold blood on the roadside. Bringing the violence home, in October 1984, one of my uncles, who had survived genocidal violence during the Partition of India in 1947, while on his way to buy groceries, was blown into bits by a bomb planted by the Khalistanis.

The Golden Temple complex had become a sanctuary for murderers and criminals. To prevent police raids, Shabeg Singh, a retired Indian army general and a close associate of Bhindranwale, had strategically fortified the temple, filling it with a huge arsenal. Enemies were kidnapped and brought to the temple complex for ransom, torture, and execution. With Punjab reeling under terror, the Indian government had finally decided to act. The government sent in the Indian army to flush out the militants from the Golden Temple. The blundering assault, termed Operation Blue Star, was launched on June 6, 1984. The army generals arrogantly presumed an easy cakewalk against the militants. They were quickly humbled. Secure behind fortifications, equipped with machine guns, anti-tank missiles, and rocket launchers, the well-entrenched militants inflicted heavy casualties. The army lost seventy-nine soldiers in the first hour of the assault. It took the army three days to overcome the militants and clear out the temple complex. Shelling by the tanks severely damaged the Golden Temple's most sacred build-ing, the Akal Takht. Hundreds of militants and innocent pilgrims trapped in the premises were dead. Operation Blue Star was a di-saster and alienated the entire Sikh community. Even those Sikhs opposed to militants turned against the Indian government. Bru-tality and violence by the militants increased exponentially after the Blue Star. By the time the violence came to an end, about 25,000 people were dead.

Amidst this backdrop, Raju had disappeared without a trace in late October 1984. I did not know if he was dead, in hiding, or worse, planning his next move. Six months after Raju left Hisar, I had received the news about him from K. S. Sidhu. After receiv-ing Sidhu's letter, I learned that Sidhu was a well-known terrorist incarcerated in Canada.

The story of K. S. Sidhu had begun at the Vancouver Interna-tional Airport in Canada. On January 9, 1985, Canadian immigra-tion officer Gail Marie Kafal was at her station in the international arrivals area. Passengers from Japan Airlines Flight 012 had started trudging through customs after an exhausting nine-hour-long flight from Tokyo. Kafal watched as a slightly built East Indian man approached her. He handed her a Swiss passport identifying him as Nikolaus Andreas Nuscheler. Nuscheler was born in Switzerland and had lived in India for several years. He had ten dollars with him and was visiting his brother, a Swiss Air employee. He planned to travel Canada with his brother. Kafal examined the passport; she noticed that the seals on the page did not match the seal on the photograph. The passport was altered. It seemed that the original photograph had been removed and a different one placed there instead. She escorted the man to an interview room.

"This is not your passport," she told the man.

"No, this is my passport," he insisted.

"No, it is not your passport. I need you to tell me your true identity. I want to know the truth or you're going to have to discuss this with the police."

Sidhu relented.

"My name is Kulwant Singh Sidhu. I purchased this passport in India from an individual named Kumar."

Sidhu had paid 10,000 Indian rupees for the passport. Kumar had affixed Sidhu's photograph to the passport. Kafal passed the investigation to the Royal Canadian Mounted Police (RCMP). Officer Colin George Abel arrested Sidhu for traveling on a forged passport.

The first place Sidhu contacted was a gurudwara (Sikh temple) in Vancouver. Shortly thereafter, Sidhu received visits from Khalistan movement sympathizers and radicals; Canadian Khalistani supporters had known about the pending arrival of Sidhu in Vancouver.

Canadian Khalistan supporters raised substantial sums for Sidhu and hired the famed John Taylor and Sons, Vancouver, British Columbia, for Sidhu's legal representation. John Russell Taylor was a well-known immigration attorney; he also had been a member of parliament.

During his trial, Sidhu testified that he was associated with

Sikh terrorists and the Indian police were looking to arrest him; if arrested, he feared execution by the police. Sidhu's parents had died in a car accident in Delhi in 1967. He was adopted and brought up by the renegade Indian army general Shabeg Singh. Shabeg Singh, along with Bhindranwale, had been killed during the army assault on the Golden Temple.

Sidhu was unabashedly proud of his terrorist ties, boasting that he was well trained in guerrilla tactics. He admitted to doing gunrunning for Sikh militants. On May 13, 1985, the judge found Sidhu guilty of entering Canada on a forged passport, sentencing him to the time served. Sidhu was placed in detention, pending his political asylum request. A known terrorist fugitive, accused of murder, gunrunning, and blowing up trains, was seeking political asylum in Canada.

———————

A FEW WEEKS BEFORE SIDHU was found guilty in Vancouver of using a forged passport to enter Canada, I was visiting Jaipur. It had been almost a year since Raju had fled Jaipur and five months after he had disappeared from Hisar. My parents were still traumatized and humiliated by what Raju had done. Although there was nothing I could do to help with the shame and social ostracism, my presence reassured them.

Late in the evening, I was reading in my bed when I heard a loud knock. As I opened the door, I was pushed aside and four men rushed inside, surrounding me. An olive-colored police jeep had blocked the front gate. The house was surrounded by cops. One of them glowered at me and then smiled contemptuously. "So I found you!" He was Inspector Sharma, Crime Branch, Delhi Police.

"What? Who are you?" Dumbfounded, I was shaking in fear. "What are you talking about?"

"I have been following you for four months. You stayed just one step ahead of me in Delhi."

"What is going on? I don't understand," my father said to Inspector Sharma. "My son has done nothing. He is just visiting for a few days."

Sharma inspected me carefully. "Who are you?"

And the police finally discovered that Raju had a younger brother!

"Why didn't you tell the police that Raju has a younger brother?" Sharma snarled at my father.

"I have talked to them countless times over the last year and have answered everything they asked me. What more can I do?"

Inspector Sharma had been after Raju since November, after Raju had stolen the Swiss passport.

"Did he come and stay with you?" Sharma asked me.

"No, he didn't."

Sharma didn't show his disappointment. He asked my father and me to come to the local police station next morning. "If you don't come, we will come and drag you there," threatened one of the men in uniform.

Next morning, we showed up at the police station. It was my first experience with the police. Sharma began his interrogation.

"How can he commit a crime and hide from us for a year?" Sharma sneered. "I can't believe that he could do anything without your help." As if Raju couldn't have committed crimes without our support. We were responsible and had to pay for his crimes. "I am going to put a twenty-four-hour uniformed guard in front of your house. You won't be able to show your face to the neighbors," Sharma threatened.

"We are already buried in shame," VK told Sharma. "How could we be humiliated any more than we already are? No one talks to us because of what our son has done."

"You should put a big reward for his capture," said Sharma. I wondered if Sharma was asking us for a fat bribe if we wanted to leave the police station. Such shakedowns were the standard operating procedure for the Indian police. I looked at my father helplessly. Sharma had to leave to answer a telephone call, giving us a short reprieve. When he returned, he looked chastened.

"You have very high connections. You can go now."

Hoping against hope, just before we left to see Inspector Sharma, VK had called Kanwar Dhir, a distant relative, to ask for help. Kanwar was an inspector with the Rajasthan Police and had recently assumed command of the neighboring police station. In the middle of our interrogation, Sharma had received a call from Kanwar. Whatever Kanwar said to Sharma allowed us to leave without hours of harassment and paying a ransom for our release.

WHILE SIDHU WAITED FOR A decision on his asylum appeal, he remained in custody at the Lower Mainland Regional Correctional Centre in suburban Burnaby. On August 20, 1985, an immigration adjudicator decided to release Sidhu if friends or supporters came forward with $2,000 in cash and $5,000 in property as a surety. This unleashed a diplomatic firestorm. India's consul general in Vancouver, Jagdish Sharma, expressed shock that a Sikh separatist, wanted in India on murder charges, was being granted bail.

Three days later, the Canadian immigration department reversed the bail decision. Al Thiessen, the regional immigration manager, issued a statement that the reversal was based on "new information received about the alleged activities of Mr. Sidhu in India." But the information had been available for months. Sidhu had admitted to his ties to terrorism in front of a judge. Additionally, the Indian government had asked Canada several times to deport Sidhu for "murdering a policeman and for an attempt to blow up a train."[*]

Denying bail to Sidhu did not mean he couldn't get political asylum. Canadian Khalistanis had been operating effectively in Canada for years, helping many militants from Punjab move to Canada after Operation Blue Star. The Canadian politicians, reckoning that being hard on Khalistani militancy might alienate their Sikh voters, had ignored the overt signs of militancy.

The Indian government's efforts to have Sidhu deported from Canada were not having much of an effect. The Indian consul general continued his media campaign against Sidhu. Canadian radicals and Khalistan sympathizers continued to visit Sidhu; Sidhu was a hero to the Khalistanis. At the same time, the slow wheels of the immigration agency continued to move forward; the self-proclaimed terrorist fugitive was about to have his asylum application accepted. Then, in February 1986, when the process was almost complete, Sidhu dropped a bombshell. He called the immigration judge and asked for a voluntary deportation. "Why would you want to do that?" the judge was surprised. "Your application is almost approved!" The judge recommended that Sidhu wait for the asylum decision. Sidhu did not change his mind. He wanted to be sent back to India.

On March 3, 1986, Sidhu was escorted by a security detail to Delhi, via Tokyo, and released in the custody of the Indian police. Japan Airline Flight 491 from Tokyo arrived at 8:30 A.M. Crime Branch investigators from the Delhi police were waiting. He was taken to police headquarters. A judge approved fourteen days of police remand for Sidhu. In theory, a remand is to allow the police to investigate the case, interrogate the accused, and obtain evidence; in reality, it is treated as an opportunity for torture and forced confessions. Torture and beatings of the accused are so

---

[*] "Canada asked to deport Sikh." Times of India, June 2, 1985, page 9.

common that the Indian courts do not accept confessions made under police custody! During the interrogation, Sidhu admitted to twenty murders.

———

TWO WEEKS AFTER SIDHU WAS deported, I was in Jaipur for the spring break. March 21, 1986, was a clear, pleasant day, with the blue sky stretched across the horizon like a taut blue sheet. My parents were at work. Sitting on our rooftop, I was soaking in the sun, reading, and enjoying the views of the Aravalli Range. I heard someone banging hard on the front door. We rarely had visitors. Who could be visiting at this time? I walked down to the front door.

Nothing begins or ends well when the Indian police are at the front door. I stood facing Inspector Sharma. He could read the shock and dismay on my face.

"I knew you were lying to me last year," Sharma sneered as he stepped in. Neither he nor his partner was in uniform. What new predicament did I face now? I wondered.

"Step out with us," Sharma ordered. I silently followed.

A white Ambassador was parked in front of our house. Two of the occupants in the rear were in uniform. They sat surrounding a stocky man dressed in a dark blue pinstriped suit.

"Look inside. Do you know who this man is?" Sharma asked.

I peered inside the car. The man in the pinstripe suit was Raju. He had gained weight and looked large and muscular. Even his complexion had turned lighter. Dumbfounded, I stared in astonishment until Raju grinned and extended his hand. His arms were shackled with thick iron chains. Smiling and looking confident, he gave me a knowing look and said, "I have told him everything."

That did not explain anything. What did he tell Inspector Sharma that I was supposed to know?

"I have already taken him to Hisar, Jind, and everywhere he

had been," Sharma told me. So Sharma knew Raju had been hiding in my dorm two years ago. What does this have to do with me? I wondered.

"You don't know how lucky you all are! The home minister had ordered your entire family to be arrested. People wouldn't have seen you again," Sharma said.

What could Raju have possibly done that India's top law enforcement official would want us all to be arrested? What was going on? Was I in big trouble? My mind was racing. Sharma interrupted my train of thought.

"He will be at the Gandhi Nagar police station tonight. You can see him there," Sharma said. "Tomorrow I will take him back to Delhi. Come and see me tomorrow morning at Ashok Nagar police station."

"Okay," I said feebly.

Sharma looked me in my eyes and said, "You can thank me for saving your whole family." He looked serious.

———

RAJU'S CRIME SPREE HAD BEGUN with his split from Amar Singh. I did not know what had truly happened between Amar Singh and Raju, except Raju's side of the story. What Raju did after disappearing from Hisar was horrifyingly shocking. He had played a game with the devil. Not only did he use the forged passport to flee to Canada, he also assumed the identity of a notorious Khalistani terrorist. The dreaded terrorist Sidhu, arrested at the Vancouver airport with the forged Swiss passport, was none other than Raju! How he managed to con a cold-blooded group of mass murderers in Punjab and terrorist sympathizers in Canada for almost a year was a mystery.

In late October 1984, after leaving Hisar, Raju was staying at the YMCA hostel in Delhi, planning his escape from India.

Hanging around on the balcony of the first floor, he overheard some clean-shaven young men speaking in Punjabi. They were Khalistani militants who had fled from Punjab, waiting to join other militants. They were trying to connect with a fugitive they had never met before. The person they were waiting for was K. S. Sidhu.

The exodus of Khalistani terrorists who had escaped army action had continued from Punjab after Operation Blue Star. To avoid detection, they had gotten rid of their turbans, cut their hair short, and shaved their beards. Using forged passports and fake names, the Khalistanis were trying to flee India. New Delhi had become a gathering point, as most of the embassies and consulates were based there. Raju had struck up a conversation with the militants, pretending to be a Sikh who was wanted by the police. He told the young men that he was K. S. Sidhu and had escaped the army assault on the Golden Temple in June, that he was a close associate of General Shabeg Singh, that he was responsible for transporting illicit arms to the Golden Temple. The naive young Khalistanis, expert murderers but easy victims for a glib Raju, were easy dupes and believed Raju's stories! They became his helpers.

Raju began networking with other Khalistanis. He traveled to Punjab and stayed with the family of a Khalistan sympathizer in Jalandhar. He smuggled weapons for the Khalistanis by crossing the India-Pakistan border on foot with three of his Khalistani handlers; they were received by a Pakistani gunrunner across the border in Pakistan. The group traveled to Lahore, where they purchased four sacks full of weapons. Their supplier transported them back near the border. The group walked across the border into India, carrying weapons of death to be used in India, courtesy of Pakistan. Who knows how many innocent human beings had their lives cut short by the guns Raju helped smuggle for the Khalistanis!

Pakistan was actively supporting the Khalistan movement. Khalistani militants were receiving sophisticated weapons and training in guerrilla warfare from Pakistan. Notorious Khalistani terrorists were being sheltered in Pakistan. Raju was offered a Pakistani passport in Lahore but declined. He was not interested in staying in Pakistan; his dream had always been to live in the West.

Maintaining the illusion that he was a major terrorist, working with real militants, Raju had flown from India, heading for Canada on January 2, 1985. The custom officer in Delhi, a Sikh, had looked at the forged passport and silently pointed him toward the terminal. The Canadian officer was not fooled by the forgery. Perhaps her Indian counterpart was not fooled either; perhaps he sympathized with the Khalistanis. Perhaps he had deliberately let another Sikh radical walk to freedom on an obviously forged Swiss passport.

In Vancouver, Raju took advantage of the confusion among Khalistani radicals. Many Sikh militants had been killed during and after Operation Blue Star. Many had been "disappeared" by the Punjab police, who were actively engaged in the extrajudicial execution of militants and innocent Sikhs alike; good information was hard to come by. Raju had arrived in Canada during this environment of mistrust and ignorance.

The radicals needed Sidhu, he needed the radicals, and the charade continued for over a year. Then something happened. Raju asked to be deported back to India, knowing full well that he would be walking into the hands of the Indian police and get tortured; be imprisoned for the rest of his life. Something earth-shattering must have happened for Raju to leave Canada. How could it be better to be imprisoned in a dreaded Indian jail instead of living freely in Canada? He never explained why he asked the Canadian government to deport him.

————

THE SWEDISH TOURIST, Nikolaus Andreas Nuscheler, had reported his passport stolen in Delhi on October 17, 1984. By itself, the theft of a passport was a minor crime, and the case file would have disappeared in the dusty cabinets of the Delhi police. However, in November the person who stole the passport impersonated Nuscheler and stayed at the Taj Mahal hotel, one of the most expensive in Delhi, then fled without paying the bill. Even this wasn't serious enough to move the wheels of the Delhi police. What caught their attention was something more sinister. The Nuscheler impersonator had been sending and receiving messages at his room, implying he was trading in illicit weapons. Twice, in the middle of night, he hired the hotel's taxi to make long-distance overnight trips. And he had fired a pistol in his room before fleeing. The investigators did not know that the impersonator—Raju—had rented the taxi in the middle of the night to go and assassinate Amar Singh. Twice he attempted to kill Amar Singh but did not succeed.

The chief security officer of the hotel, besides reporting the incident to the police, contacted the inspector general, Crime Investigation Department, warning that whoever stayed at his hotel was involved in international terrorism and espionage and posed a threat to India.

Then the faux Nuscheler had ended up in Vancouver. The rest was history.

While the officers from the Research and Analysis Wing, the Indian CIA, were interrogating "Sidhu" and were attempting to implicate him in various terrorist incidents, Inspector Sharma was expressing his doubts. He had suspected from the beginning that Raju had stolen the Swiss passport. One of the hotel employees, who was Raju's classmate in Jaipur, claimed to have remembered that Raju had impersonated Nuscheler. But the Nuscheler impersonator had then ended up in Vancouver as Sidhu. And then that selfsame person had become the cause célèbre of the Khalistanis in Canada. Stolen foreign passports, especially European passports, were commanding high premiums from the Khalistanis fleeing India. Raju could have sold the passport to Sidhu.

Having spent months chasing Raju, Sharma instantly knew the person he went to take custody of at the airport was not Sidhu. When Raju told the top intelligence agents that he was not Sidhu but Kaushal Dutta from Jaipur, the ministry of home affairs was notified. Maybe Sidhu was impersonating Kaushal Dutta; maybe Sidhu had received help from Kaushal Dutta. Everyone related to Kaushal Dutta was to be picked up as a terrorism supporter.

It was Inspector Sharma who argued against having all of us arrested. It was indeed Sharma who saved us from brutal third-degree treatment by the police. He had solved the Swiss passport theft case and the Nuscheler impersonator case, showing that it was a con man and not a national security risk who had fled the Taj Mahal hotel after firing a handgun. Raju's long, convoluted flight from the law and his con games had finally come to an end. His anger, his quest for revenge, and his arrogant selfishness had brought him and us to near catastrophe. Had Inspector Sharma not interceded on our behalf, our entire family would have ended up in a secret prison, getting tortured or worse. What upset me the most was Raju's extraordinary cunning and resourcefulness. He was intelligent enough to fool the Khalistanis, the Royal Canadian Mounted Police, the terrorism investigators from the U.S. Department of State, and the Canadian criminal justice system. Why couldn't he use his intelligence to better his life and to help his family instead of imperiling their well-being? Why was he trying so resolutely to take revenge? If he couldn't force himself to reconcile with Amar Singh, why couldn't he just forgive him and move on? Perhaps it was too much to ask of him. My own father and his siblings were certainly not the forgiving types. They had made vendettas part of their lives. Raju did not have any positive role models influencing him to be forgiving. Neither did I!

8

MUST LOVE END IN A TRAGEDY?

Once, we had an agreement between us,
Perhaps you remember, perhaps you don't;
The promises you made, to be faithful,
Perhaps you remember, perhaps you don't!
Once, we did desire each other;
Once, we shared life's path together;
Once, we too were in love together,
Perhaps you remember, perhaps you don't!
— MOMIN

Why do great love stories end in tragedy?

Was Orpheus destined to lose Eurydice in order that Jean Cocteau could make a fascinating movie? Cleopatra and Mark Antony had to die in the end, Romeo and Juliet, Layla and Majnun, Heer and Ranjha, Anna Karenina—did they have to perish? Why did Devdas die in front of Paro's house before getting a glimpse of her face?

My love was neither epic nor legendary. Was it going to end in tragedy?

Before Wes left India in June 1982, we had fantasized about our future; we wanted to spend the rest of our lives together. But there were difficult decisions, the main one: where would we live?

Wes loved India, but not enough to live there. I could not even imagine leaving my birthplace. Then there was the question of whether I could provide her with the comforts she was used to in America. Who would compromise, and how much?

But agreements could wait. We still had to finish college, to stand on our own feet. There was also that other elephant in the room—Steve Booth, the long-distance American boyfriend with whom she wanted to end up before she had met me. Soon I was going to be her long-distance boyfriend, too. She did not want to leave me; she was not willing to let Steve go, either. Besides, "It is too early for me to decide to marry," she had said.

I did not force Wes to make a choice between Steve and me. "I don't care what you do for the next four years. Just come back to me after we finish college."

It was a blatant lie. I was perhaps the most jealous and possessive man in the world. I wanted no man to ever get a chance to touch her. But why should I fear or be jealous? The Sufi poet Mian Muhammad Baksh has said that "a heart permeated with true love for the Beloved has eyes for nobody else." I was in love, I did not have eyes for anybody else; if she loved me, she couldn't have eyes for anyone else.

Insanity provides a great defense against reality; that's why true lovers must be insane. I was close to being insane, but unfortunately not quite there. I understood the challenges I faced. I knew the American way of life was more open, more permissive, less reverent. She had fallen in love with me, forgetting about Steve, with whom she had previously wanted to spend her life. How long would it take before she would forget me in lieu of someone else? Nevertheless, I was ready to sacrifice my jealousy, my desires, my being, and to wait for her to return.

I had resolved not to abandon love, to let whatever must happen, happen. I would not expect her to return to me, but I would not give up on her, either.

I kept baring my soul to her in letters. The responsibility for expressing the depths of my devotion was borne by paper and ink, carrying my message of love and longing across the vast oceans. Perhaps my passionate poetic love letters did the trick, perhaps distance had not lessened her love, and perhaps reunion was not just a dream. Wes was equally in love with me. She missed me and still wanted to spend her life with me; she wanted to come and visit whenever she could afford it. We continued our long-distance romance on the wings of aerograms. There was hope. There was a possibility of reunion.

———

THREE YEARS AFTER SHE LEFT, in July 1985, the beginning of my senior year at Hisar, I received a long letter from her. I was overjoyed. Wes was coming to see me in September and would stay for four months! The long dark dismal night of separation was finally about to end. She would be here in less than two months! I excitedly read the details of her flight to Delhi. Then I turned to the second page.

"I love you so much. It's been a torture to be away from you," she said. "I have been desperately trying to visit you the last three years. Finally, I have a chance. I have wanted to be with you for so long."

Words can be hollow and feckless. How could I trust anything she said? In her long letter filled with her declarations of love for me, of her desire to be with me, only one line stood out for me: "I am seeing Jonathan Post. We are having a serious relationship. I will not have sex with you."

The letter fell to the floor, pages scattering. I slowly dropped to my knees, holding my head. After what seemed like an eternity, I managed to struggle up; I started punching the walls until my knuckles were bloody. Everything within and without had gone numb. I walked to my closet and pulled out my sole material possession of any worth: a stack of her letters and a pile of our photographs. I started to shred the letters, not stopping until every single one was ripped to bits.

Once I had destroyed her letters, I stared at the stack of our photographs: pictures of two ecstatic teenagers, affectionate, inseparable, loving, smiling, kissing, passionately embracing, holding hands, looking soulfully into each other's eyes . . .

I torched the remains of the letters and the pictures. A small fire, the funeral pyre of love, smoldered near my mattress. I then did the unthinkable—I took her lovingly preserved lock of hair from the envelope and tossed it on the fire. I had completely lost hope; I had lost everything.

Acrid smoke from the pile filled the room. Cards, letters, pictures, reminiscences, mementos—everything was gone; nothing remained but ashes.

Love is destined to end in tragedy, so better now than later!

The blow came from out of nowhere, knocking me down. I was in shock. There was no forewarning. Had she misled me over these past three years? Did she lie when she said that she loved me? That she wished to spend her life with me? What else did she lie about?

If she was done with me, couldn't she at least break the news in a kinder fashion? Is this how life-transforming intimate news is shared? If she had hidden things from me for so long, why couldn't she just wait two more months and tell me in person that she had a new lover, that we were finished?

In two lines, she had destroyed our relationship. She could not have been crueler if she had tried.

In that hot, muggy, and suffocating room, with my dreams and hopes strewn over the floor in a pile of ash, I wrote the reply.

"Wes," the letter began. No "dear" or "my darling"—she didn't need to read any further to know my reaction.

"Perhaps you have some reason to visit India, I don't know what that reason is. You say that you want to see me. I don't know what that means. I will not see you if you come to India. If you still decide to come, that is your choice. I will have my friends help you find a place in Jaipur, but I will not see you. I *cannot* see you."

I could not understand her reasoning. Her letter bluntly told me that we were finished, that she belonged to someone else, that the intimacy between us was history. At the same time, how could she claim that she was coming to India because she loved and missed me? Why did she say that she wanted to be with *me* for four months? What for? What was she trying to accomplish by visiting India after her letter? What was *wrong* with her? *I* was supposed to be the insane one in our relationship.

She did receive my letter. She did not cancel her trip. I was heartbroken. I was thoroughly confused. I could not understand what she was trying to accomplish. There was nothing left of our relationship except false, broken promises. Wasn't she finished with trampling on my heart?

———

TWO MONTHS LATER, I WENT to pick her up at the New Delhi international airport. She had cruelly stabbed me in my heart, yet I was not angry at her. Instead, I was concerned for her welfare: she was coming to India by herself. I felt responsible for her. It was insane, but I had decided to take her to Jaipur, find her a place, and leave.

On August 22, 1985, late in the evening, I stood waiting at the passenger arrival area at the New Delhi airport. I was confused, questioning my judgment, questioning why I came to help her. I did not want to see her, yet I was looking forward to seeing her. Sufi poet Hazrat Shah Niyaz had said that in love, one must drink from the cup of selflessness and oblivion, without worrying about the outcome. *Liberate yourself from rationality, let whatever must happen,*

happen! My decision was made: I had liberated self from rationality, come what may.

I saw her step out of the terminal. She had lost weight and looked frail; her hair was longer. Last time I saw her, my face had no facial hair; I sported a virgin mustache and a beard. Our eyes met.

I took her in my arms; she hid her face in my chest. We held each other tightly, oblivious to the glares of puritans, oblivious to those who had to dodge around us, oblivious to everything. We were silent. We had no words; the tight embrace and the silence were expressive enough.

With her arms still tightly wrapped around me, I held her face in my palms and stared in her eyes, as I had done numerous times before, trying to find my old lost love. We smiled. I gently kissed her forehead.

We took the midnight bus from Delhi to Jaipur. She rested her head on my shoulder the entire way; I wrapped my arm around her waist and kept my hand in hers. By the time we arrived in Jaipur at five A.M., my arms and shoulder were numb; it was a pleasure; she could have rested her head on my shoulder forever.

So many things had changed in three years. My parents had always been happy to see her the last time she was in Jaipur. Now both were opposed to her visit. In July, I had informed VK about her four-month-long visit.

"Tell her not to come. If she still wants, tell her to visit for two weeks and leave!"

Her visit was an annoyance to them, almost a threat. During her last visit, she was with her family, she was going to school, and she was learning Kathak dance, which made her less threatening, more acceptable, more honorable. The situation was different now: an unmarried young white woman was visiting their sole remaining son. Her visit, in their opinion, had no other purpose except to be with me. People would be scandalized; they would talk, ask difficult questions. Weren't people already scandalized because of what Raju had done?

"I don't get to tell her how long she stays," I had retorted.

"Then let her rent someplace and stay away from here. We don't want to see her."

I was surprised and angry. I did not want to see her, but she had been my beloved, my friend; how could I stop being protective of her? Who would look after her during her stay? I would be an overnight train ride away from her. I would be concerned for her all the time.

If Wes was going to stay in Jaipur for four months, for her safety and comfort, I preferred that she stay with my parents. But they had turned against her for some reason; they were hostile to the idea of her visit, so much so that they recruited Munna Lal to convince me she should not visit India.

"Sunil, it is not safe for you or your family's peace. Raju has police cases going on; every other day cops are here, making life difficult for your parents. If she comes here, the police might create more trouble."

It was the only time I ever lost my temper with Munna Lal. "What the hell are you talking about? What does she have to do with Raju's crimes or police harassment?"

————

WHEN WE REACHED JAIPUR, Wes was exhausted after the twenty-hour-long trip from the United States and the subsequent five-hour-long jarring bus ride. She did not pick up on the subliminal hostility of my parents; instead she was happy to see them.

Next day, we found a place for Wes at the house of Mr. P. P. Keswani, VK's colleague. It was a private studio apartment attached to Mr. Keswani's home. Ironically, it was but one block away from where Wes and her family had stayed three years ago. I had passed by the house many times on my way to her home.

Wes was pleased to rent a place of her own, allaying my fear that she would feel unwelcome because my parents did not invite her to stay with them. In the end, it turned out to be for the best that she had her own place.

We walked to our old haunt, the Raja Park market, and purchased household things. I helped set up her little kitchen and her new home in Jaipur. The day flew by as we fixed her little apartment; suddenly it was late in the evening and I had to leave.

We had spent the entire day together, yet we were a million miles apart. We did not hold hands or hug each other; we did not kiss; we did not even look into each other's eyes. The realization that we would never be intimate left a deep painful void in my heart. I did not know how she felt. I dared not ask.

I gave her a wary hug before leaving. "I love you," she whispered, holding me tightly; she did not want to let me go. I was quiet; what could I say?

I grimly walked back home through the dark empty streets. In three short years, starting with Raju's flight from the law and ending at Wes's July letter, my life had turned into hell.

Next morning, Wes rode a bicycle I had lent her over to our house for breakfast. After breakfast, I walked her back. Unlike before, we did not ride together on the bicycle. Every moment with her was a sad reminder of the past.

I had to leave for Hisar in the evening; I stayed with her for the few hours before my train departed. It was miserable: two wary ex-lovers awkwardly sharing the privacy of a quiet room, keeping their distance. What could be more heartbreaking?

For reasons known best to her, Wes started detailing her experiences over the last three years, telling me everything except what I wanted to hear. She did not apologize or make any effort to soothe my hurt feelings; instead, she decided to break my heart with the stories of her affairs. She talked about Jonathan, about others with whom she had had sex without any serious relationship.

"Why are you doing this to me? Have you no heart? You have already destroyed me! Don't tell me anything. I don't want to hear any of this. I don't want to know anything."

She did not stop.

She told me about every single one of her dalliances since leaving Jaipur, going over each in detail. I may have been provincial and narrow-minded, but is there any lover in the world who would enjoy listening to such tales? This is what she was doing *while* leading me on for the last three years, insisting that she loved me and wanted to spend her life with me?

"Stop!" I turned away from her, feeling defeated.

Devdas died before he could get a glimpse of Paro. At least Paro was not a wanton woman; and if she was, thankfully, she did not reveal it to Devdas; it would have made Devdas's death that much more agonizing.

Wes came up behind me and held me. She was crying. "I have come to India because I love you. I want to be with you."

"I don't know what that means." I could not look at her. "What am I supposed to believe?"

"Why don't you believe that I love you?" she said, weeping.

"After all you have said and done? After you spent hours telling me about people you have slept with, having sex with people you were not even interested in? After you kept me in the dark for the last three years? After you kept leading me on as if I was the one you wanted?" I continued mournfully: "If that's what you mean by love, I don't want your love. Hate me instead."

"I am sorry. It's been very painful for me." She started to cry.

Painful for *you?*

At that moment, I should have walked out of the room and out of her life. If her callous self-serving nonanswer failed to jolt me out of my stupor, what would?

The heart is an enthusiastic purchaser of humiliation, as Ghalib would say, or maybe I was simply an incorrigible idiot who deserved every bit of the humiliation and misery she was dishing out. To be in love must mean one is able to forgive unconditionally, is absolutely insane, or is utterly delusional. I was not delusional; most likely I was not the forgiving sort, either.

I turned around and hugged her.

"Please don't cry." I wiped her tears and kissed her forehead. "I will come and see you as soon as I can. I have asked Rummy and Anuj to keep an eye on you and help you." Rummy and Anuj, my friends, lived in Tilak Nagar, not far from Mr. Keswani's house.

What a spineless craven excuse of a man I had turned out to be. I went to pick her up at the airport when I shouldn't have; I sat listening to the stories of her wanton behavior; I bared my chest to allow her to stab me; I took her abuse without complaint. And in the end, *I* was the one wiping off *her* tears; I was the one consoling her; I was promising to visit her again.

———

LATE THAT EVENING, ON MY way to Hisar, I waited for my train to depart from the Jaipur railway station. I was in despair.

The loud whistle and the clanging of bells announced the train's departure, pulling me back to the earth. As soon as I heard the engine start and the train's wheels begin to spin, I became claustrophobic and couldn't breathe. I wanted to jump out of the train and run until my lungs exploded and my legs collapsed. I wanted to keep running until I reached her. I wanted to be with her!

My stoicism was fake; I missed her deeply. Every moment I had pretended to be aloof near her, I had wanted to hold her in my arms, to never let go. My own heart had betrayed me; it yearned for my tormentor. Why was I yearning for such a cruel, unfaithful woman?

Love is either true or transactional. Desiring someone, wanting to spend your life together is transactional; Sufis call it *Ishq-e Majāzi*, worldly or physical love. *Ishq-e Haqīqi*, the true divine love, abandons desire. It comes with no regrets, rationality, doubt, fault, or pain. It requires no give-and-take; it needs no physical touch; it doesn't even require the presence of the beloved. True love longs for the union of the souls, not of the bodies. If my love was true, why should I have any regrets or pain? How could I suffer if my love was true? *Drink from the cup of selflessness and oblivion; let whatever must happen, happen!*

———

EARLY THE NEXT MORNING, physically I was back in Hisar, but my heart was left behind in Jaipur, and my mind—well, who knew where my mind was? I couldn't survive like this. I resolved to concentrate on my studies and ignore everything else—which happened to be an untrustworthy twenty-three-year-old woman who was not mine.

I should not see her, I told myself; therefore I will not see her. I resolved not to go to Jaipur until she returned to the United States.

A few days later, I was on a train leaving for Jaipur, deliberately making a mistake of monumental proportions. Nothing good was going to come out of this, my mind cautioned my heart; the heart thought otherwise—it wouldn't even let me stop at my home to see my parents. Instead, to the shock and disappointment of my parents, I went directly to her. Decades later, I found VK's comment about the incident in his diary: "This boy is acting reckless. He has become blind in love." He wasn't charitable to Wes, either: "Wes is glued to him all the time. I have told her that we Indians don't like this Western type relationship; we cannot tolerate it." A few days later, he noted: "Wes always remains sticking to Sunil whenever he is at Jaipur. We don't like it." While my parents might have been shocked and angry at my impropriety, Wes was overjoyed.

She was beaming as we embraced. As I held her head close to my heart, suddenly everything became peaceful; the raging turmoil in my mind, the conflict, despair, sadness—everything seemed to melt away.

Over the next four days Wes and I retraced our steps from before, strolling in the parks, walking in the Raja Park market, going to see romantic Hindi films, eating from the same plate, feeding each other. When we weren't out and about, we were inseparable: we cuddled in the bed, we stroked each other's hair, we smiled, we sighed.

We were together again; it was so achingly beautiful.

But I was living a miserable life of contradictions; I was sad and happy, I was in despair and at peace. I was my own enemy. I did not want to leave her side, to return to college, to be away from her. But return I must. This time Wes came to see me off at the train. She smiled, she waved, she walked with the train as it slowly moved from the platform. It was heartbreaking, but I had to smile because everyone was staring at me.

Soon I was back in Jaipur again. The college students, on strike over some trivial demands, had shut down the campus. I made a request to one of the strike organizers before leaving for Jaipur: Please do me a favor and don't break the strike for a few weeks, even if the college accepts all your demands. He stared at me quizzically, asking, "*Paagal ho gaye ho kya* [have you gone mad]?"

No one in Jaipur had been expecting me. I went straight to her home. It was dark. I gently knocked at the gate. The lights came on in her room. She peered out and noticed me standing outside the gate. I saw her break down in tears, and she came running to the gate. She was weeping and smiling, squeezing me in her arms, not letting me go. My shirt was wet with her tears. We stood in the small courtyard for what seemed like eternity. She could not stop crying.

"It is so painful to be away from you," she said. "Do you know how much I love you?"

In my head I replied: *Yes, it was painful for me, too, and I doubt if you can even imagine how much I love you.* I said nothing.

I kissed her eyes and forehead as we walked in. We slept on her bed, our bodies pressed together, her head resting on my heart, every particle of my being in turmoil. I wanted to kiss her lips, but didn't.

I had willingly given up my happiness for her. Peace and sanity had disappeared from my life. My heart was in no-man's-land.

Through the last three years, I had been hardening myself. I did not expect love in return for love or for her to be faithful; I was ready for my heart to be broken; I was prepared to be crushed—so she had obliged me! Ghalib, in his inimitable style, had said:

I can sleep well at night because I was robbed during the day.
Not worried anymore, I must thank the robber.

What I had feared, what I did not want, had happened. She had been sleeping with others. She had permanently cured me of my fears. I should have thanked her.

But she had not come to India to torment me or make confessions. She had come to India to reclaim her love. She had been serious about spending her life with me. Her statement that she would not be intimate with me did not originate with her. When she told her parents that she was going to India to be with me, Robert Bly, her stepfather, had advised her, "It will be unfair to Jonathan if you have sex with Sunil." Both of her parents preferred Jonathan over me. After all, I was an unknown quantity from a distant land, possibly a narrow-minded overbearing male chauvinist from India. I was unworthy of their daughter. She regretted being influenced by Ruth and Robert and writing the devastating letter.

———————

WHEN WES MAKES UP HER mind, nothing can stop her. I was wary, suspicious, sad, and afraid of intimacy with her; all of this did not matter to her. She would not give up; she was going to convince me of her love and devotion. She did not know, but my resistance was weak to begin with.

Within a month, despite everything, she had broken the barriers between us. Either she won me over or I caved in—the result was the same. We were crazily in love again; we could not stay apart even for a moment; once again, the external world ceased to exist. We were lost in love.

I spent more time in Jaipur than at my college. I even began sleeping at her apartment. In a close-knit old-fashioned neighborhood, an unmarried Indian man was living openly with an unmarried white woman. We were scandalizing the world.

And one day when I least expected, she dropped a bombshell.

We were riding on a scooter, going to the Shaheed Bhagat Singh Park near Raja Park, one of our favorite spots. Wes sat behind, pressed tightly against me, arms wrapped around my waist, her face resting on my back.

"I want to tell you something," she whispered in my ear.

"What?" I turned my head sideways, to enable me to hear her over the road noise.

"I want us to get married!"

"What? What did you say?"

It was the moment I had dreamed about, a moment I thought would never happen. This could not be true. I must have been in a dream; and if I was, I did not want to wake up. Oh god, please, don't ever let me wake up!

When we reached the park, the petunias, zinnias, daisies, marigolds, periwinkles, jasmines, frangipani, China roses, and honeysuckles were in full bloom. We sat near a row of flowers. I held her hand. She was beaming, but as soon as I looked in her eyes, she shrank back, turning red.

"It is too painful to be away from you!" she said wistfully. "I can't live like this anymore. I love you so much."

"You already know how I feel," I said, fighting the urge to hold her in my arms; we were not in the right place. I was struggling against tears.

My devotion, my unconditional love had won; she wanted to spend her life with me. My love would not become a tragedy. I will not end up becoming Devdas! *Love had arrived, this world and the next one had been achieved, my body of dirt had risen to the heavens!*

That afternoon we became secretly engaged. We went to a local shop, the Bakewell Bakery, to buy a cake. Back home, we took turns feeding each other.

"Could you please come and join me in New York?" she asked.

To join her, I will have to sever my ties to my birthplace, family, culture, and friends—in effect, cut myself off from everything I know. My parents had been made refugees due to the legacy of colonialism, hatred, and religious nationalism. I would become a refugee again if I left for America, a willing refugee of love.

"I will do anything for you."

———

EVERY DAY WE SPENT TOGETHER was blissful. We were at peace, with the knowledge that we had each other; conflicts, pain, anguish, everything was left behind in the past. But her four-month visit was coming to an end; she had to return to United States. On December 17, 1985, I went to see her off at Delhi. Bidding farewell to a loved one is never easy; however, this time my heart was not mired in despair. When she turned to wave goodbye, I so wanted to leave with her. But I would be seeing my sweetheart in just a few months. I had survived three years of separation; I should be able to survive seven more months! And this time, we will never be pulled apart again.

As I walked out of the airport, I recalled Rumi's couplet:

Rationality ruins this world.
Love saves the world and the hereafter.

———————

SANITY AND RATIONALITY WOULD HAVE meant the end of our relationship months ago. Soon insane irrational love was going to unite us once more thousands of miles away.

9

FLIGHT OF LOVE
ACROSS THE OCEANS

I lived on your promise, though I knew it was false.
For if I had believed it, I would have died of happiness.
— GHALIB

Our days and nights of togetherness were suddenly over. Once again being apart from her tormented me. But this time, separation would be ending in reunion.

She was finishing up her bachelor's at New York University, I at Haryana Agricultural University. She was going to school and working to save money for my migration; she was busy figuring out how to bring me to the United States. Every few days I received her letters. Our separation pained her as well; she sounded more desperate than I. "Please come soon, I cannot live without you," she entreated. "I am dying without you. I cannot wait anymore."

Young single Indian males, unless very wealthy and privileged, had a ghost of a chance to get a visa for the United States. This of course meant that I could never see her unless she wished to live in India. But she was determined to find a way to bring me to the United States.

"I found a visa category to bring you here. It is called a fiancé visa. We will have to marry within three months of your coming here." We had not planned on getting married so soon; however, neither of us was opposed to the sole requirement of the visa.

She filed the petition in New York City.

By the end of April, her petition was approved. I was ecstatic; now I could go to the American embassy in Delhi and apply for the visa.

But I was fearful. Three months after Wes returned to the United States, Raju was deported from Canada as Sidhu. When my fiancé petition was approved in New York City, Raju was still in police custody; the elite Crime Investigation Department was still interrogating him. For a year and a half, Raju had been in the company of militant separatists and terrorists in Canada. What secrets did he know? What crimes had he committed? The security agencies were determined to pry open the vaults and find what was hidden inside. Sukhvinder Singh, a Sikh in Haryana who had sheltered Raju, had been hauled in by the police. I had no idea how many others were being harassed and tormented by the police because of their association with Raju.

The possibility that the police might arrest me because of Raju, ruining any hope of getting a visa to the United States, was too real. I went to my cousin Kanwar Dhir for advice. Everything about Kanwar is large—his belly, his mustache, his heart, his laughter. His cheerful disposition rubs off on others. His eyes twinkle and he laughs easily at small things. Serendipitously, he happened to be the new commander of the police station investigating Raju's bank forgery case. Kanwar was the rare relative who had not shunned us after Raju's crimes; he had been very helpful.

"Kanwar, I am afraid that the police will arrest me. Raju stayed with me at Hisar. He also forged the Swiss passport there. What should I do? Should I get anticipatory bail? I cannot afford to have a police record; I will never get a visa for America. The American embassy has already asked me for a police clearance certificate!"

"*Mustt raho!*" Kanwar smiled. "Don't fret over anything, just relax. You don't need any anticipatory bail. Just tell me how that girl of yours is doing?"

"She is doing well. She keeps asking me to come to America right away!"

"Make preparations to go to America and start a new life. You have nothing to worry about," Kanwar said.

A heavy weight was off my shoulder. That day, to the surprise of my parents, I revealed I was leaving for America. "Wes is sponsoring me for immigration to America. I will find a job there." I did not discuss our secret engagement and our upcoming marriage. I did not have to; everyone knew what was going on.

———————

A MONTH EARLIER, WHEN HE moved Raju to Jaipur after his deportation, Inspector Sharma had allowed me to see him at the police lockup. Two years after Raju had left Hisar, we faced each other across the bars in the jail. When he disappeared, he was skeletally thin with straw-colored bleached hair. Now he was astonishingly large and muscular. His hair was dark, his skin was light; there was no evidence that the prison had damaged him.

The situation was full of irony. I had always thought that Raju was superior to me in every way; that day he was locked in a cage like an animal. Yet he looked calm, almost unconcerned at his situation, as if everything was going to sort itself out—almost as if soon he would be in control of his fate again. That confidence was the essence of Raju. Standing on the other side of the partition, I was the one racked by uncertainty.

Not knowing how to begin the conversation, I stammered out: "I thought people lose weight in a prison!"

He laughed. "The jail in Canada is better than how we live here! I ate a lot of meat and lifted weights."

"Do you realize how much trouble we have been through because of you?"

He nodded. It was too dark inside to read his expressions.

"No one is helping us. All our relatives have cut off relations with us. Our parents want to bail you out. The problem is, if you escape, they will be ruined financially. What do you want us to do?"

Our parents did not have the money to bail him out; they would have to put up the house as surety. If he defaulted and became a fugitive, our parents could lose the home they lived in.

"I will not do anything. I will not run away!" he said. Probably he would have promised me anything to be bailed out. Could we trust him? I did not know. Nevertheless, I recommended to VK that he should get Raju bailed out. Despite my parents' best efforts, bailing out Raju took five months. He was released from prison on August 14. He was back home after his self-imposed exile, while I was preparing myself to go into exile; he had taken refuge in a foreign country to escape from the law, while I was going to a foreign country to become a refugee of love.

In June, Wes had sent me a check. There was enough money to purchase my ticket to the United States and an emerald set in a 22-karat gold ring for our wedding. I could finally leave for America. But it was not to be. It seemed we were doomed; everything had ground to a halt. The visa application could not be processed without a letter of financial support. Wes was a student with no assets and little income; her letter of financial support was not sufficient. Someone in a better financial state needed to submit a letter of support for me. She called Ruth, her mother, who lived with her husband, Robert Bly, in Minnesota, for help. The discussion did not go well.

Wes's mother went on and on about all the flaws in our plan: *This is crazy. How could you even think of bringing someone from India and getting married? This is a bad decision. You are too young, you are not ready, you haven't finished school, we are paying for your college, you have so many outstanding student loans, neither of you have jobs. How are you going to live? You will end up in poverty for the rest of your lives. What about your different cultures and ethnicities? How will you cope as an interracial couple with people's hostility? Don't be crazy.*

Another round of discussion followed, more concerns were raised, more attempts were made to get Wes to change her mind and act sane, to forget about summoning that unknown Indian man to the United States, to give up the idea of getting married to him. Robert and Ruth refused to sign a letter of support. And then everyone Wes knew joined the army of naysayers.

Indian men beat their wives. They want absolute control. They want too many children. They are not good to women. You will live a horrible life. Indian men leave their wives and return to India. Every possible stereotype was trotted out. Pressed from all sides, Wes was becoming concerned. But she was determined, she was adamant. We are engaged! He is the only one I want to spend my life with, she insisted. I *will* bring him here. I don't want to marry anyone else. No one could change her mind.

She called David Ray, her adoptive father, with whom she had visited India in 1981. Unlike others, David was neither upset nor condescending. He sounded supportive, but he refused to help. To sign a letter of support meant accepting my financial responsibility for three years; he was not prepared for that.

A letter of support stood between our union.

Wes decided to approach Jim Perrin, her biological father. He was her last hope. Her stepfather and adoptive father had already said no, so she prepared a list of reasons to convince Jim. Jim knew that Wes had dated a man in India.

"I want your help to write a sponsorship letter for Sunil. Robert and David have already refused to help. I am trying to bring him here on a fiancé visa. Let me start with why you should . . ."

Jim refused to listen to the reasons he should provide a sponsorship letter. He stopped Wes before she could read him the list.

"You don't have to tell me any of that," Jim said. "I can certainly find it in my heart to sign a letter of support. You don't need to tell me the pros or cons!"

Jim was delighted to hear about our marriage. "You should have lots of children," he told Wes.

Jim had never met me, he knew almost nothing about me; yet without any hesitation, he helped his daughter bring her Indian fiancé to America.

The letter of support arrived in the mail on August 5, 1986; on August 6, early in the morning, I stood in line in front of the U.S. embassy in Delhi. Tense and worried, I hoped that the visa authorities would believe my unbelievable story of meeting an American teenager in Jaipur and falling in love.

Dozens of visa seekers were ushered in a waiting room that was quiet as a graveyard and sterile as a hospital lobby. After a long, agonizing wait, I was called in for an interview. A middle-aged Indian man in a woolen sweater stared at me across the plastic safety barrier. He was not friendly.

"How are you, sir," I greeted.

He vacantly stared through me, without responding.

"Who is she? How do you know her?" He pointed at my application.

I handed him a bundle of photographs, copies of the photos that I had destroyed one fine July afternoon last year. These photos had survived because Wes had a copy; without these, no one would believe in our fairy-tale romance.

"What is this?" I could feel his hostility across the thick partition.

"These are our photographs. We met in Jaipur five years ago. She visited again last year."

The interviewer started flipping through the photographs; there were dozens of them. He suddenly stopped.

With disgust plastered on his face, he sneered at me. "*Accha, vahan jaane se pehle hi American ban gaye?*" The level of contempt he projected could only be conveyed in our native language. *I see—you have already turned into an American,* he had said. He might as well have added *debauched* in his condescending remark. The reason for his revulsion: he was looking at a picture of me kissing a beaming Wes on the cheek.

"You fucking asshole!" I muttered under my breath. I wanted to punch through the bulletproof barrier and slap his smug face.

He pulled out Jim's letter of support from the application.

"Where is the original? This is a copy."

"This is what I received from her."

He did not ask any more questions, he did not say a word, he did not acknowledge my presence. But in the end, it did not matter. I was handed back my passport containing a K-1 visa stamp; I finally had the fiancé visa in my hands. I was elated. A new life beckoned far away in a foreign land, where my beloved lived. The fairy-tale romance was ending in a reunion. Love had not ended in tragedy.

———

THE DAY AFTER GETTING THE visa, I returned to Jaipur. A week later, Raju walked out of the prison, his bail was finally approved. One son had returned home; another son was leaving home.

I was booked to fly from Delhi on September 4, 1986, on an early-morning Pan Am flight. In less than four weeks, I would be in New York City. I was going to join the woman I had yearned for every single day of my life since I had met her. I should have been ecstatic and blissful, but I was afraid.

I called Wes to give my flight details. "Are you going to be there to pick me up on time?" I joked. "Or am I going to end up staying at the airport for the day?" Wes was always late.

Despite my attempt at humor, I felt distant; something was wrong. What could it be? Was I developing cold feet? Was I concerned about moving to a far-flung corner of the world, with no control over my life? Perhaps the fact that neither Wes nor I had a job or a way to support us was bothering me. I couldn't put a finger on what the problem was. It felt as if Wes was cold toward me, too.

Two days before my flight, my parents and Raju took me to Delhi. It was an inauspicious beginning to my journey. A pall of terror hung over the city. Khalistani terror attacks had spread from Punjab to Delhi. On September 2, bombs had been discovered in Delhi public transport buses, just before they were to explode. The city was crawling with police. The usually bustling streets were deserted in the late evenings. On the night of my departure, we were stopped twice by the police and our taxi searched. I looked at Raju as the cops shined their lights on us and searched the car; was this a sign for Raju to lead a straight life?

In the midst of the crowded airport, I was feeling intensely lonely. I was saying goodbye to the life I knew, embarking on a new journey; in the no-man's-land between the two, I stood, confused, disconcerted.

My smile had disappeared when the time came to bid goodbye. VK could see I was distressed. He patted me on the back. "Enjoy your life and don't worry about anything." My mother stood by quietly. Raju looked cheerful.

"You have to take care of the parents now. Don't do anything to hurt them again," I told Raju. I turned away, each step bringing me closer to my beloved.

ON SEPTEMBER 4, 1986, before dawn, I stepped inside a Pan Am DC-10 jumbo jet, Flight 4986, bound for New York City via Frankfurt. I looked in awe at the long rows of seats, stretched ten across. I had never been inside a plane. After nine hours, the plane landed at Frankfurt. I stepped out to stretch and made a quick call to Wes; she was not home.

The next leg of the journey was peaceful. Most of the passengers had disembarked at Frankfurt, so the plane was mostly empty. We had left Delhi at five A.M.; due to the time difference of nine and a half hours between Delhi and New York, I was in New York City in the early afternoon of the same day. As I walked toward customs, I noticed the first difference. In Jaipur, Wes looked different from everyone else; here, I was the one who looked different. Compared to India, everything was clean, the people were all professional, everything was nicer, everything seemed to work. I stepped in front of the immigration officer, a young blonde in a crisp uniform. For what seemed like an eternity, she kept examining my passport; then she walked away from her booth, my passport in her hand.

What the heck was going on? What was wrong? She returned with her superior, a tall, large-bellied man with thinning blond hair. Perhaps she needed help because she did not know how to process my visa; maybe the immigration personnel rarely came across fiancé visas.

Her superior looked down at me. "Who is going to pick you up?"

"My fiancée."

"Show me a picture of your fiancée."

"I don't have a picture of her on me. All of our pictures are in my checked suitcase."

"Are you trying to tell me you don't carry a picture of your own fiancée in your wallet?" He gave me a skeptical look. I was prepared for them to grill me about how we had met, where we had met, but I had never thought that they would demand to see her picture.

"How can I carry a photograph of her in my wallet? I don't even carry a wallet!"

Please, please, don't send me back to India.

"I can show you dozens of our photographs if I could get to my suitcase."

The man retreated to the booth of the blond officer and gave her instructions; I strained to hear him without success. Minutes later, the officer handed my passport back.

"Here you go. Good luck!" she said.

The wait of my life was finally over. We were not thousands of miles apart; she was waiting on the other side of the wall. In a few moments, I could gaze at her, hold her in my arms; within moments, I could kiss her.

I stood in the arrival area, scanning the crowd, searching for her. There she was! Something was wrong; she looked subdued. We embraced; it was over too quickly. What was happening? Was she being unfriendly at the moment of our reunion? Maybe she was tired or stressed out; my arrival had been a weight on her shoulders for months. She had struggled very hard to bring me to the United States. She was going to school full time; she was working long hours to save money for us. Now she could relax—*we* could relax.

"I am hungry, I need to eat," Wes said. I quietly watched her eat a hot dog. She did seem to have something weighing on her mind.

"I have to move today. The place I live in is too small. We needed a bigger place for us."

So we were moving, the same day I arrived from India.

From the airport, we went to the studio apartment she shared with a Dutch woman. In the back of the taxi, I pulled her close and kissed her. She looked shy, hesitant, unsure. "It was a nice kiss," she said. I was shocked. My beloved made this hollow, insipid remark when I kissed her? *A nice kiss?* Something was seriously wrong.

We arrived at her studio, which was indeed microscopic—the bathtub was in the middle of the living area! How could people live in such cramped places? I wondered. The mover was already there, waiting with his small pickup truck. The next few hours were a blur: out of the taxi, into the studio, move boxes to the truck, get en route to the new apartment. Jetlagged, wiped out, my mind a jumble, I held on to her hand for comfort.

We arrived at a tower block on Montgomery Street. She had found a beautiful apartment on the twentieth floor; it was large, with windows opening in three different directions. We were going to share it with its genial Chinese owner; we had our own large private room.

Helping the mover cart boxes from the truck to our room took some time.

When we were last together in Jaipur, I had helped set up her apartment. One year later, we were setting up a place of our own. By the time we had finished piling the boxes against the wall, I was wiped out. I collapsed in the bed.

Wes came and sat next to me. "I have to tell you something!"

I held her hands and smiled. "What is it?"

"I am pregnant."

"What?"

"I was having a relationship. I am pregnant."

Her hands slipped out of my hands.

"How could you do this? How could you do this to me? How could you do this to *us*?"

I stepped away from her, walked to the window, and stared down at the ground. Would it hurt when I hit the ground?

"I was lonely!"

Could she have chosen a worse response? I exploded in rage.

"How the fuck can you say that? You were having sex with someone while planning our wedding, while you were trying to bring me here? How could you betray me like this?"

"You were so far away. I was feeling lonely."

"You are beyond pathetic. You don't even have an *excuse* for what you have done. What does it tell us about you?"

This was my first day in America. A few hours ago, I had stepped into this country on a fiancé visa!

"I love you. Please don't leave me," she said somberly.

"You love me? Who on this planet can trust you anymore? Do you even understand what you have done? You have betrayed my absolute love and devotion for you. You destroyed our loving bond."

I opened my suitcase, removed the box containing the wedding ring, took it out, and tossed it out of the window.

"No!" she screamed, and ran toward the window.

The ring struck the side jamb and fell on the windowsill. Wes grabbed the ring as I tried to pull it out of her hands. She tried to save the ring, holding it next to her heart.

"Please don't do that. It's my wedding ring. Please don't throw it." She was crying.

I stepped back. I was still stunned. How could she have sex with someone while calling me here, while pledging her devotion and love, while planning our wedding? Did she mislead and lie to me or to the other man who made her pregnant? Did she warn him that her "relationship" was going to last only until her fiancé arrived? Or did she make promises of eternal love to him?

The "day of reunion" had turned out to be the worst day in my life. The ringing of the telephone pulled me out of my sulk. Wes picked up the phone. I could hear a male voice on the other end; she mostly listened. It was close to midnight. On September 4, 1986, in New York City, I stood an arm's length from her, discovering that my devoted fiancée, who claimed she could not live without me, was not yet done with the man she supposedly had an affair with.

"I know." "I understand." "I know how you feel."

Was it possible to degrade or dehumanize me anymore? What

could she do to make me feel more worthless? Could our relationship be cheapened any further? Could I fall any lower?

She walked to me and thrust the telephone in my hands.

Oh, how I wished I were a masochist. This would have been the most intensely pleasurable experience in my life!

The man on the other end of the line sounded miserable.

"Please! Wes and I love each other too much. Please don't come between us! Please leave us alone. Please don't kill our baby!"

"What?" I uttered in shock.

"Why are you coming between us? Go away. We love each other. Don't hurt us . . ." he was still babbling as I tossed the phone aside.

What the fuck? When did I ever come between two lovers or try to sabotage their relationship? When did I try to kill their unborn child? Go for it! Live happily ever after with your child, I am not stopping you. *I* am not the one who would ever stand between two lovers or sabotage their relationship.

That night we slept on opposite sides of the bed. I felt numb—numb to pain, numb to feelings, numb to my humanity. Lost, confused, I slowly drifted into sleep.

———

EARLY THE NEXT MORNING I woke up to the sound of the phone ringing.

Ruth was calling from Minneapolis.

"Mom wants to talk to you."

"I don't want to talk to anyone."

She shoved the phone in my hands. "Please talk to her."

"Hello, Sunil! Welcome! How was your flight? How do you feel in your new place?"

"Thank you. I am fine." I mumbled. What else could I tell her?

"Robert and I are looking forward to seeing you. We are looking forward to the wedding."

Looking forward to the wedding? What would you say if you knew your daughter was pregnant by someone else?

She knew! While I listened to Wes discuss our trip to Minneapolis in two weeks, I learned that the trip to have an abortion had already been scheduled. I felt like a protagonist in a Franz Kafka story.

Empires collapse, civilizations disappear, rivers change their course, but life still goes on. Wes had to go to school. Perhaps she was afraid of what I might do alone, so she asked me to accompany her. We took a short subway ride to New York University. I sat in her class and listened to a lecture on U.S. foreign policy. An hour later, we walked to Washington Square Park and sat on a park bench.

I held her hands and asked, "Why did you do this?"

She started crying.

Shedding tears is a woman's ultimate weapon; women use tears to protect themselves from a hard male-dominated world but also to defeat the most formidable of their opponents, to melt the hardest of the hearts, to overcome resistance, to conquer. Women deploy tears to be endearing, to show love, to beg forgiveness, to manipulate, to evade difficult situations, to dodge responsibility . . .

I took her in my arms as she sobbed.

"You are the only one I have wanted to spend my life with. I love you. Please don't leave me. You promised me you would be the best husband in the world! You promised me you would love me forever!"

"I don't know if promises have any meanings."

"Don't say that. You promised me you would never leave me. Please don't. You *know* how much I love you." She continued to weep.

I held her tightly as she buried her face in my chest.

Her betrayal was overwhelmingly colossal; yet she was the same woman I had given my heart and soul to. I still cared for her, I still loved her, I couldn't see her unhappy. I wanted an explanation, a rationale—in fact, *any* rationale for what she had done. Maybe something momentous, something critical, would explain her behavior and vindicate her. If there was nothing else, she could simply say that she was sorry, that she had slipped. I would have forgiven her. But love without forgiveness cannot be love. It is reduced to give-and-take, a transaction. It is merely physical, earthly love— *Ishq-e Majāzi*. *Ishq-e Haqīqi*, does not require any explanations; it requires acceptance and absolute forgiveness. My rationality was at war with love's insanity.

I held her hand on the subway ride back. We walked to our apartment as I tried to absorb the shock of my new surroundings, the plethora of new sensations, new sights, new smells, new notions, new emotions, new people, a new culture, a new language, a new way of communicating—an entirely new way of life. I was in America, the land of dreams and opportunity. My dreams were already dead.

"Two days ago I was in India, looking forward to spending my life with you. Two days later, here I am, walking with my fiancée, who is pregnant!" I looked at her pensively. While I thought only great love stories ended in tragedy, the love story of a nonentity like me had become a pathetic tragedy as well. I had become Devdas.

A WEEK HAD GONE BY in New York; we were spending every moment glued to each other. I accompanied her to her NYU class in the morning and then to her Kathak dance class in the afternoon. We spent the rest of the day walking the bustling streets of the city, hand in hand, arms around each other. We cuddled, she sat on my lap, I looked in her eyes, we smiled, we looked like lovebirds.

Back home, I argued with her. Every time we kissed and made up, we nestled in each other's arms, we slept holding each other close. I did not fight with her because I wanted to. My upbringing, my conditioning, my Punjabi heritage required that I make a show of my disapproval. Part of my mind believed that I was required to object.

Whenever I look at the pictures our Chinese roommate took

of us shortly after my arrival to America, I notice the joy exuding from our contented faces, the beaming young couple blissfully lost in togetherness! The Urdu poet Seemab had eloquently expressed my feelings: *In my passion to kiss the roses, I lay my tongue on sharp thorns.* Regardless of what she had done, I was happy to be with her. I had achieved my life's goal! True love cannot die or have regrets. Pain and tears alone could reveal the preciousness of love. In spite of her faithlessness and betrayal, my love for her was unshaken. Despite threatening to leave her, I never *once* contemplated leaving her. The crack in my heart and mind did not go away, but it was clear to me that nothing mattered if we were together. The anger subsided, only to be replaced by gratitude. I could be a Devdas *and* I could get my Paro at the same time! As a Sufi poet once said, *we have surrendered to the religion of love; we have sacrificed our life for love.* I was not the type to abandon love, else I would have walked away from her last year.

Despite slipping and falling, despite the blunder she made, she was passionately in love with me and wanted to spend her life with me. Despite her fears, despite the attempts people made to dissuade her, she had not given up and had been determined to bring me here; against all the odds, she did manage to do it. We were united because of her steadfastness.

On November 28, 1986, we were married in Moose Lake, a small Minnesota town. I finally met the rest of her family, including her three fathers, five stepsisters, five stepbrothers, grandmothers, and cousins, at a small Lutheran church, for our wedding. Jim Perrin stood with me near a pew, watching the wedding rehearsal. He pointed accusingly toward David and told me, "Here is the man who stole my wife." Jim was happily married; he had three daughters, a son who was a mirror image of Jim, and another child on the way, but he had not forgotten his first love. Who does? We smiled at each other. He was proud to see the eldest of his three daughters getting married.

10

SCIENCE, IDEALISM, DISILLUSION, AND THE SEARCH FOR MEANING

Where is the second step of desire, O Lord?
In this immense world of endless possibilities,
All I see is a single footprint!
—GHALIB

After our marriage, Wes still had a semester left to graduate. New York City was not a place for someone with my background—a person with no work experience and a degree in horticulture. In India, those who could afford it went to school and sought work after graduation. The rest were relegated to menial positions. There was nothing in the middle.

I struggled for months to find a job to support us until the Central Park Conservancy hired me to take care of the grounds. I worked with a crew who cared for Strawberry Fields, the bowling greens, and the massive turf area. But my heart was not in it. All my life I had wanted to do something useful for society: fight for truth, help the helpless. I was convinced that scientists were nobly doing research to help humanity. I wanted to be a scientist. I also wanted to leave New York City. Exciting and colorful, with a fantastic nightlife, New York is a city that never sleeps. But I hated crowds and large cities. Wes was the sole reason I was here. She did not want to leave. She loved the life in a fast-moving city, and New York offered many opportunities for her as an actress and a dancer.

Wes was a strong confident woman, not a pushover; she strived toward her goals. Yet despite her interests, despite her desire to be an actor, after her graduation in the summer of 1987, she gave it all up for me. She made the sacrifice for us to be together. She sidelined her career in acting and agreed to follow me to Gainesville, Florida, where I began my graduate studies and research in plant physiology. She might have been trying to prove her devotion to me or compensating for our starting off on the wrong foot; whatever the reason, she was completely supportive. She had become even more loving than before, as if that was possible; she was a dream partner. We were "always stuck together like Velcro," as Maury, one of her dance troupe colleagues, used to say.

WHILE I WAS WORKING ON my master's degree, Raju worked in a hotel in Jaipur, teaching French on the side. He was on the road to reform. His criminal case in Delhi was delayed; the prosecution was unable to proceed without critical evidence from Canada— which the Canadian government was slow or hesitant to share. His criminal case in Jaipur was also off to a slow start; the police had not finished their "investigation." The cops had spent an inordinate amount of time harassing us and shaking down my parents for money; I joke that if the police here had put an equivalent number of hours into solving the Kennedy murder mystery, they would have solved it by now.

The courts in India work at a snail's pace; sometimes the defendant, the witnesses, or the victim literally die of old age before a trial is complete.* Not much has changed in the Indian criminal justice system. A large part of it remains a criminal system devoid of justice. This meant there was hope for Raju; he could stay out of prison for a long spell as the creaky wheels of justice moved ever so slowly. He was intelligent, outgoing, and resourceful; maybe he could turn things around.

Two years after I left India, Raju displayed his resourcefulness. He wished to attend an annual seminar for French teachers in France, an all-expense-paid trip with the cost covered by the French government. Raju submitted his application to the French embassy in Delhi. He was selected for the fifteen-day-long seminar in Paris. But it was an exercise in futility: he was barred from traveling. The police had seized his passport because the prosecution thought Raju was a flight risk.

How Raju convinced the judge that he could be relied upon to travel abroad still remains a mystery. The judge ordered the police to release Raju's confiscated passport and authorized the visit to France. Within days after reaching France, Raju called VK. He had met a woman, Christine, in the hotel he was staying in. He had stayed at her house for two days. He wanted to marry her! Raju was not someone who would fall in love at first sight. He was calculating about everything, always thinking about whether he could derive benefit from something or not. Women were not part of the equation; he generally looked down on them as inferior and subservient to men. Yet within a week of his arrival in Paris, he had convinced a Frenchwoman to marry him. He returned to India after the seminar. The passport office once again confiscated his passport.

In Jaipur, Raju sought the permission of the judge to go and get

---

* *Raju's criminal case in Jaipur was finally adjudicated in 2006, twenty-two years after the reported crime. The court declared him an absconder. He was charged with forgery, and his trial and sentencing were postponed pending his apprehension.*

married in France. This time the judge was skeptical. Why would he return to India to face trial and imprisonment after marrying a Frenchwoman? Wouldn't he just stay in France, avoiding the possibility of a long prison term in India? The court rejected the petition. Raju appealed to the Rajasthan High Court. Surprisingly, this court approved the petition; he could travel to France for his marriage, with the stipulation that he must return in two months.

This time Raju did not return; it would have been foolish of him to go back to India. Raju's fleeing India this time was the responsibility of the court, shielding our parents from financial ruin—they did not have to pay the bond. Twice a judge had allowed Raju to travel during the trial. Raju would never go back to India again—to return meant forfeiting his bail; being prosecuted for additional crimes, including being a fugitive; and facing a stiff prison sentence. The Amar Singh forgery case alone could have put him in jail for twenty years. Once again, he had become a fugitive.

Raju had cast a spell over the Frenchwoman, Christine, and her family. They barely knew him, yet they took him in; he lived off them for some time, until he was allowed to work. Employing his incredible powers of persuasion, he could convince people to do anything.

Three years after I had seen him last, he proudly called me from France, when his son was born, in 1989. As he had always wanted, he was living abroad; he was married and now had a child. He could never return to India, but that did not concern him; he now had an opportunity to create his own nest in France. I breathed a sigh of relief. My parents were disappointed. Parents were supposed to live with and be supported in their old age by their eldest son, but that possibility had evaporated. At the same time, they no longer worried that he faced long imprisonment at the conclusion of his trial. He had engaged in deadly con games, seemingly without any repercussions. Life had given him an opportunity to begin afresh, I was hopeful he would take it.

I also began the next phase of my life, moving from the University of Florida to the University of California at Davis. I was the first one in our clan to pursue doctoral studies. In May 1989, I moved to Davis to work as a postgraduate researcher in plant biology, before starting my PhD in August. Davis was a wonderful change from hot and muggy Florida. Wes and I felt more at home in California. We had easy access to the great redwoods, the Sierra Nevadas, and the beaches of California. Wes was very pleased; California provided opportunities for her to pursue her acting and dance.

AS A CHILD, I BELIEVED that scientists helped change the world for the better. In addition to my contrarian attitude, I was an unapologetic idealist; I wanted to make a difference, to live a purposeful life. While completing my doctorate, I mentored several inner-city teenagers; I also supported many organizations, some focused on poverty, some on police brutality, others on various aspects of social justice.

Although I was involved in social activism, the reality of street gangs, drive-by shootings, and inner-city problems did not affect my life. I was becoming an academic, living in an ivory tower. I loved my research into the ways sugar molecules joined together to create strands stronger than steel and incorporated themselves into the cell walls of plants, protecting them from pests and diseases; the factors that cause a fruit to ripen but also initiate the steps toward cell death; the manner in which certain plant hormones receive signals from the environment and regulate senescence or seed germination or defense responses against predators. These were important areas of research and it was the right time to be in science—the technology was advancing at a rapid pace, providing the means to answer fundamental questions. Life was good, and I was intent on spending my life in laboratories.

I was on the way to becoming a research biologist. Then one day I realized that academic research was becoming unhinged from basic science. The focus of a scientist was on getting funding, publishing papers, writing grant proposals, and populating a lab with intelligent, highly skilled researchers who were paid almost slave wages; the university had turned corporate. I still remember my first major funding proposal, seeking government funding for our lab.

My research proposal was accepted, prevailing over the requests for funding of other more established scientists. We received funds to conduct research for two years. I had managed to convince hardened scholars that my research topic was convincing enough to be funded. I should have been proud! Instead, I was troubled. In the proposal, as expected, I had cited the potential benefits to society if our lab received funding. It was understood that the benefits were greatly exaggerated, to the point of being untrue. Everyone made similarly extreme claims to get funding to run their laboratories! The day I received the notification of the award, I started questioning what I was doing, wondering if I had the stomach to mislead the grant agencies every time I submitted a funding proposal. I started questioning the career I had chosen.

I was still in the middle of my working on my doctorate when Wes finally decided to take the plunge and focus completely on what she had always wanted to do: acting. The mecca for actors, Los Angeles, was four hundred miles south of Davis. She made the decision to move to Los Angeles, leaving me behind in Davis. I preferred the small university town to the sprawling, ugly, dangerous land of gangs, drive-by shootings, and earthquakes. I couldn't even dream of moving to Tinseltown. I wished that she wouldn't leave. But she had followed me faithfully for six years, so how could I deny her dreams?

Soon after she moved, Southern California exploded in rage. The trial of four white LAPD officers accused of beating the black motorist Rodney King—the first ever video in the world that had gone viral, long before YouTube came into existence—had ended in acquittal for the officers. Riots had broken out in Los Angeles. Wes was there all by herself. I was frightened and miserable because I was not there to protect her. If anything happened to her, I would have been devastated. Fortunately, while the city burned, she remained safe.

ON APRIL 29, 1992, an all-white jury was set to decide whether the officers accused of using excessive force on Rodney King, a black motorist, were guilty or not; *everyone* expected the officers to be found guilty, based upon the video of the incident. Less than nine hundred LAPD officers were guarding a city of millions spread over almost five hundred square miles. A few farsighted officers warned that the city needed to be prepared for potential civil disturbance after the verdict; their warnings were blithely ignored. The stage was set for a devastating conflagration.

At 3:15, the verdict was announced. Much to the shock of everyone, the officers were acquitted of using excessive force. Within minutes after the verdict was read, crowds began to form in South-Central Los Angeles.

The area south of downtown Los Angles had been in economic decline for decades. It was awash in violent crime and a hotbed of gang activity. One out of four South-Central residents was on welfare. Racial tensions were simmering between the blacks and the Latinos and the blacks and the Koreans. As the unrest over the acquittal of the cops spilled into the streets, angry gang members were the first to join the uproar. The first incident involved the looting of a Korean liquor store by black gang members.

The famous intersection at Florence and Normandie was overrun by hundreds of enraged residents throwing rocks and bottles. The violence of the scene was too great for the officers who were called to the scene. The beleaguered cops faced angry confrontational mobs. At 5:43 P.M., Lieutenant Michael Moulin, the watch commander of the 77th Street Station of the LAPD, fearing for the safety of his outnumbered officers, told his officers to retreat. It would turn out to be one of the worst policing decisions ever made.

Flying rocks and bottles and the echo of gunshots followed the retreat by the police. Mob violence erupted while the oblivious arrogant chief of police Daryl F. Gates attended a political fundraiser in wealthy Brentwood.

Asian, Latino, and white motorists driving through the area were attacked. Gang members drunk on looted booze pulled motorists out of their cars and beat them mercilessly. News helicopter hovered over the violent mobs, transmitting horrifying images of Reginald Denny's being beaten nearly to death. Everyone in the world could see the mayhem and looting in real time. They could also see the police had abandoned South-Central Los Angeles, allowing the violent mobs to do as they pleased.

Within four hours of the jury's decision, smoke was bellowing out of South-Central Los Angeles. Arsonists were torching shops and buildings. Firefighters were helpless; they were confronting hostile crowds, mob attacks, and gunfire. As night fell, fifty fires burned in Los Angeles. Acrid smoke drifted over the city, adding to the pollution-saturated atmosphere of Smog City.

The world-famous LAPD had lost control. Its managers made blunder after blunder; the worst one was not to deploy officers in an overwhelming number at the initial stages of the violence, when the riots could have been quickly controlled. The conflagration that started at Florence and Normandie spread unchecked, and the city burned uncontrollably for days, until the military was called in. On May 4, when the rioting finally subsided, 54 people had lost their lives, over 2,300 had been treated in emergency rooms, and property worth a billion dollars had been destroyed.

Like millions of people around the world, I had been transfixed by the leaping orange flames and dense dark smoke over Los Angeles. I had watched in disbelief as the violence and protests spread from Los Angeles to San Francisco, Las Vegas, Denver—all across the country. The facade of a civilized society seemed to be peeling off in America, the deep racial divide made visible by the response of black America to the Rodney King verdict. In the middle of my working on my PhD, I contemplated joining the LAPD and teaching them how to be humane. A liberal joining the LAPD: it was an outrageous and fantastic thought.

On November 24, 1997, early in the morning, before sunrise, I stood outside the LAPD Ahmanson Recruit Training Center, waiting nervously with over seventy recruits, ready to be initiated into a way of life radically different from mine.

11

THE FIRST DAY

The horse of life is galloping; we'll never know the stopping place.
Our hands are not touching the reins, nor our feet the stirrups.
— GHALIB

June 4, 1998, was the day of reckoning. I stepped out of the "campus," as the seasoned street cops derisively referred to the police academy, ready to be initiated into the mean streets of Los Angeles. No more mock scenarios, pretend traffic stops holding fake aluminum pistols, and wrestling on the mats with classmates. I would be the cop in training, feeling unsure of myself and doubtful of my abilities, lacking knowledge of the ways of the streets, questioning my own resolve, strength, courage, and purpose. I wondered if criminals also lacked resolve and sometimes questioned their abilities.

I pulled into the parking lot and lifted my *war bag*, which contained forms, cheat sheets, extra ammunition, and other supplies, out of the trunk; I noticed that I was shaking. My heart was beating fast. I was sweating. Anxiety had my stomach in knots. I was about to begin a new life in an alien environment. Everything and anything I did would be foreign to me. I was thirty-four years old. Seven months ago I lived in Cedarville, where crime was virtually nonexistent and people were friendly; Los Angeles was the antithesis of Cedarville.

I used my 999 key, which let me in the back door of the station. I could go to any police station in the city and use this key to enter. The thought made me feel special, and I smiled again. Flickering fluorescent lights illuminated the drab corridor leading to the men's locker room, intensifying the gloomy atmosphere. Why are our government buildings so dreary? Was there a conspiracy to create a corps of depressed government employees in the United States? I wondered.

I passed the property room to my right, a secure area used to store contraband and evidence obtained from arrests and crime scenes. The roll-call room was to my left, the locker room to my right. I stepped into the locker room, lugging my heavy war bag. The dank, dismal room was lined with gray metal lockers, with narrow wooden benches separating these metallic coffins in which cops stashed everything: uniforms, duty belts, boots, extra ammunition, amulets, posters of naked women, pictures of guns and fast cars, quotes by their heroes, and pictures of their wives and children next to macabre photos of bleeding and dead suspects. Towels hung on bent metal hangers that poked through holes in the lockers; old shoes and slippers littered the ground; plastic dry cleaner wraps lay stuffed in the litter bins; the smell of sweaty undergarments and wet towels permeated the air. This was what the interior of a garbage bin must feel like. Maybe cops are callous and brutish because they have such a horrible workplace environment.

I was early; there was no one else in the room as I walked to my locker. Rookies, called *boots*, had to look impeccable, arrive early for the roll call, and sit quietly in the front row of the roll-call room as seniors sitting in the back rows razzed them.

It was too narrow to sit facing my locker to shine my boots, so I straddled the bench separating the two rows of lockers. I twisted a white rag tightly around my index and middle fingers, dabbed it on the shoe polish, and started rubbing the tightly stretched rag on my boots, in between sprinkling the leather with sprays of water. An officer must look sharp; spit-shining the shoes was a daily ritual. My focus on spit-shining my boots was suddenly broken. A corpulent man in uniform walked toward me, holding rolled-up sheets of pages that looked like copies of an arrest report. The locker-room floor trembled under his feet. I obsequiously stepped over the bench to make way. He slid past me, stopping at the next locker.

"Good morning, sir. How are you doing?" I forced a feeble smile on my anxious face and tried to ingratiate myself with someone I had taken an instant dislike to. The man reacted to my pro forma question as though I had tossed a lit flare in the dry Southern California mountains.

"Fuck, man, I am pissed. Can you believe this shit? That fucking judge threw out the case on a small technicality. We had the asshole wrapped up with a confession. Just one technicality in the arrest report, *one fucking sentence*, and he throws away the whole shit. He let the motherfucker out of jail, free to rob and kill again. Why the fuck am I wasting my time putting assholes in jail? I should just pick them up and drop them near that motherfucker's home. I would love to see daughters of those motherfucking liberal ACLU assholes get raped; maybe *then* they will stop loving the criminals. They are fucking up our country. Fucking liberal college boys, they have never been to the real world. Living in their fucking fantasy land." He spat out words mixed with spittle.

I cringed in fear. Had I been transported to a different universe? The trainers at the police academy were right. Things were indeed different in the "street" than on the sterile campus of the LAPD Police Academy. Nobody cursed in the academy, there were no diatribes against "liberals," and some of the officers in the academy even seemed politically progressive.

This was a different universe altogether. Was *real* training about to begin?

I wondered if this man's attack was directed at me. I was told that officers would get the lowdown on every newcomer to their division. But how would this guy, whose name I later learned was Brett, know that I was one of the dreaded *college boys* he despised?

He towered over my six-foot frame and weighed maybe about three hundred pounds. Compared to other LAPD cops I had met in the academy, who seemed obsessed with the shape they were in and their appearance, he was a relic. He paused, gave me a hard look, and asked, "Hey, boot, *you* a college boy?"

"No, sir, not at all," I lied, surprising myself. It took no effort or thought to lie, but it also did not stop me from being deeply ashamed. Minutes inside a police station and I had lost my courage and integrity. I was lying about who and what I was. Where was all this going to lead? Where was it going to end?

My heart pounded. I wondered if Brett could read through my anxiety. I also wondered whether an ACLU liberal was worse than a criminal or just an equally despicable creature to Brett and his cohort. Would he prefer me over my own brother who was a criminal? Or did he hate criminals and liberals equally?

I shivered as I wiped beads of sweat from my shaved head. I exhaled slowly, my shame tinged with relief. I was afraid to tell the truth to this crusty old-timer. The lie saved me from a contrived explanation. There was no fast way to explain why I joined the LAPD after spending years in academia. Would these hardened cops believe what I would tell them anyway? Would Brett believe that my idealism led me to abandon a safe life and the prospect of leading research teams in high-paying industry jobs in order to join the LAPD? That I wished to wash away the stains of shame and dishonor that my brother had brought to our family? That I was tired of the rampant materialism in my life and was looking for a way out through public service?

My decision to join the LAPD was complex. I was guided by idealism, shame, and a sense of failure. I wanted to spend my life helping people as a scientist. And yet by the mid-1990s, I had lost my idealistic notions about science. After all, it didn't seem to be solving the human problems I wanted to address; instead of serving humanity, science was serving the profit-oriented large corporations. I also wished to wash away the stains of shame and dishonor that my brother had inflicted on our family. My experience with the Indian police had made police corruption and brutality all too real to me. I had also become heavily influenced by Punjabi Sufi poets and earnestly believed the precepts they propounded about abandoning a life of desire and rampant materialism. I surmised that as a police officer, I could touch the lives of victims, live my idealism, and also confront the notorious system from within.

I had no faith in my ability to face down hardened career criminals. I was not physically imposing and could not speak the language of the streets. And I did not know if I could control them as other cops did. I was a study in lack of confidence; I was unused to violence and to the idea of using force to solve problems. Moreover, I was worried how my fellow officers might respond to my social and political views. But all we have is one life to live! I needed to take this chance.

I interviewed with the LAPD early in 1996. There were three interviewers in the small room, including two officers. The two officers in the interview room were trim, athletic, skeptical, and confrontational. They went after me as soon as the bell rang, unleashing a barrage of questions. *Why the police? You have a doctorate! Why aren't you applying to be a scientist? Why don't you work for a high-tech company making tons of money? How would you deal with the streets? Do you know it is like stepping into a raging fire? How have you prepared yourself for this job? You understand that this is not a laboratory? Have you ever been in a physical confrontation? Do you know that police work is dealing nonstop with conflicts? How would you handle conflicts on the streets?* I had no idea if they were concerned for my safety or were trying to dissuade me.

The aggressive questioning made me wonder if they would buy my pitch about idealism and public service, my desire to make a change and to be a role model. The sole civilian member of the interview team seemed amused by the list of civic organizations I supported. Despite the grilling I received, they gave me a near perfect score, guaranteeing that I would be let into the LAPD.

A few months later, Officer Stan Nelson, who investigated my background, knocked on my door. "Sunil," he said, "I am very impressed by your background. If you stay in, you should rise through the ranks quickly. We need people like you to make this department better." I don't know if he was sincere or simply indulging in salesmanship, but his gentleness and his words were helpful, they gave me courage to take the leap into a new life.

———

I TURNED MY ATTENTION BACK to spit-shining, hoping that my focus on my boots would let this officer leave me alone.

The last time I was as nervous was five years earlier, during my hours-long inquisition by five University of California professors intent on probing if I was sufficiently qualified to earn a doctorate in plant biology. Compared to facing this cop, that was a piece of cake. Even before Brett had entered the locker room, I had been feeling like a cat that had blundered into a backyard full of snarling pit bulls. Now I wanted to dash out of the locker room, jump into my car, and drive straight back home.

My concerned friends and relatives, shocked at my decision to quit academia and take this apparently suicidal leap—joining the notorious LAPD—had started making bets about how long I would last in my new career. Nobody gave me more than a week. If Wes was afraid of my new career choice, she did not show it.

Now, sitting on a splintery bench, I wondered if I would last an hour. All my idealism had run headlong into the Jell-O belly and crudeness of an old-timer. I was racked with self-doubt as I started gearing up for my first shift as a rookie street cop.

The thick front and rear panels of my bulletproof vest weighed a ton when I picked them up. The blue Velcro strips made a jarring sound as I joined and adjusted the panels to fit my torso, overlapping the front and rear ballistic panels to keep a bullet from finding its way in between. As I slowly pulled the sleeves of my dark blue wool uniform shirt over the vest, I grew even more anxious. Acid rose into my throat. *What are you doing here, you idiot? Run away while you still have time. There is no shame in leaving.* Teaching young students in Santa Monica College would be a lot more fun.

By now two more rookies had joined me in the locker room. Regular cops didn't show up until a few minutes before the roll call; Brett was there only because he had returned from a court appearance. We were there an hour early, to shine our gear and groom ourselves to perfection.

I strapped on my heavy boots and carefully slid my five-shot backup revolver in the left pocket of my uniform trousers.

"Why are ya puttin' on this full-sleeve shirt? You wanna die in the Valley heat? Do you know how hot it gets here, beach boy?" A booming bass voice emerged from another giant hovering over me, a mass of muscle standing six feet five, still wet from his shower, covered in a white cotton vest, a towel wrapped around his lower half. He was an intimidating sight—someone you would like to be on *your* side in a fight.

"Yes, sir. I mean, no, sir," I mumbled. I could see him smirking as he walked away. It was a tradition that rookies must wear long-sleeved wool shirts and ties until their training officers deigned to allow the far more practical short-sleeved shirt. In the Valley, when summer temperatures often went over a hundred degrees, wearing the thick LAPD blue shirts was nothing less than torture. But boots must never complain. They must keep their heads down and speak only when spoken to.

————•————

BEFORE I JOINED THE FORCE, I had not known a single police officer, I did not know how a typical street cop spent his shift, and I knew almost nothing about police work. In my world, cops were heavy-handed ignorant provincial slaves to the rich. My preconception of cops as ignorant muscle-bound brutes had not been helpful, for I was neither brutish nor strong.

How would I survive on the streets? Academy instructional videos and re-creations of true incidents with gruesome fights between hardened criminals and officers had exposed me to the reality that cops had to sometimes deal with reckless and dangerous humans and take risks no sane person would, for *any* price. The day of reckoning had now arrived.

What I might encounter on the mean streets of Los Angeles was only a small part of my worry. I had spent many sleepless nights wondering if I would find acceptance in the LAPD. Would I be in perpetual conflict with everyone? Would my brown skin and ethnicity set me apart, cause me to be marked and shunned or worse? And if so, how would I cope? I could not disguise my contrarian nature, ethnicity, character, and beliefs. I was probably going to be the lone "communist pinko liberal" in the middle of true red-meat conservatives.

How was that going to play out with my partners and other cops?

My reasons to join were contrarian, too. Besides my deeply personal reasons—my brother's criminal past and a desire to seek redemption for my family's evil history—I also wanted to find out if the cops routinely beat minority men and brutalized inner-city residents. I wanted to stand against what I perceived as police cor-

ruption and brutality. I wanted to be a role model as a compassion-
ate human being, as a *servant* of the people, victims and criminals
alike. I wanted to stand up for my beliefs, with pride. How would
my partners and mentors react to my philosophy, the partners who
must watch my back and put their lives on the line to protect me?
My brief encounter with Brett had shaken my confidence. Would
I have the courage to stop my training officer if I saw him beating a
black teenager?

Hypothetical questions are hard on sensitive souls. In the
Bhagavad Gita, Lord Krishna advised the warrior Arjuna to go to
battle without reflecting on the consequences—not to think about
the results, just to do what you must. Karma alone is your duty
and faith. I agreed more with the recommendation of philosopher
J. Krishnamurti: live as an observer of the world without wanting
to change it. The change will come as we honestly look at things
around ourselves, because the life of an honest observer must transform. Or would it?

———————

AFTER LYING TO BRETT, I found it easier to lie again during my ten
seconds in front of the watch. I introduced myself as an ex-teacher.
Why start on a wrong foot by saying, "I have a PhD in biology
from the University of California and a master's from the University of Florida. I received my bachelor's degree in India. I special-
ize in host-pathogen interactions, carbohydrate chemistry, and
enzyme-substrate relationships. I was a scientist . . ." The LAPD
cops believed that University of California campuses were hotbeds
of communism; why become easy prey to their sharp wisecracks? I
wasn't naive. I was sensitive to unjust criticism and did not want to
be mocked.

After the roll call, I stood in line to get the equipment and
patrol car assigned to me, while my training officer, Lisa, a short

and heavyset middle-aged blonde, waited. We started our pre-patrol rituals. I checked out a shotgun, an old Ithaca model like the one used in the film *L.A. Confidential*. It was probably made twenty years before I was born, but with a shotgun, age didn't matter. Officers and criminals equally respected "the tube," as it was reverently called. Several 9-millimeter bullets from a Beretta 92FS semiautomatic pistol have sometimes failed to stop a determined suspect, but a single shotgun blast *will* end the drug-fueled rampage of the biggest, baddest felon. It's the closest feeling to invincibility on this side of a barrel, unless the opponent happened to be clad in a full-body armor, firing automatic bursts from an AK-47—which is what happened at a North Hollywood robbery a few months before I joined the LAPD.

I checked out three radios, one each for Lisa and me and one for the fixed radio in the police car. I couldn't imagine how officers of yore coped without radios. Not too long ago, cops had to walk to their cars to radio for backup or get vital information. Sometimes hopping fences and chasing felons through the backyards, they had to shout to bystanders to call the police for additional help.

I checked out a camera, useful in documenting evidence of assault, battery, and especially domestic violence, as victims often refuse to cooperate with the prosecution. Some officers loved to photograph gruesome murder victims, suicides, or traffic accidents involving mutilated body parts for their personal albums. I didn't understand the logic, but I hadn't been on the job long enough.

The last item on my checkout list was a Taser: a small black plastic device with a pistol grip and a flat face into which we insert bright orange cartridges. Press the Taser trigger and a small charge launches the darts at combative suspects, penetrating clothing or skin to deliver a 50,000-volt jolt of electricity.

Loaded with necessary paraphernalia, we left the kit room, exited the rear station door, and headed to the covered parking garage with its three rows of cruisers—black-and-whites, also known as *shops.* While Lisa watched, I searched the patrol car thoroughly for contraband. Officers sometimes missed drugs or even weapons secreted by an arrestee in various hidden compartments and folds of an arrestee's body. Criminals occasionally discarded undiscovered contraband in the rear compartment of a police car, a difficult acrobatic feat, considering their hands were cuffed behind their backs and body restrained by the seat belt.

Next we put our war bags in the trunk. Heavy and hard to carry, they contained ballistic helmets; angled mirrors to find burglars in dark buildings; extra ammunition; police forms for every possible situation one may encounter in the City of Angels; street guides; envelopes to place evidence in; a first-aid pouch; and other miscellaneous essentials.

Last but not least, I conducted a five-point safety check on my shotgun, after which I strapped a pouch of extra rounds on its stock and secured it in the shotgun rack. Then Lisa started the engine, and I began typing all our pertinent info into the cruiser's computer—logging our serial numbers, assigned radios, beat designation, and cruiser number. This information, stored on a computer downtown, could be retrieved in case of emergency. After we completed these last rituals, we were ready to take radio calls and hit the road.

———

LISA WAS SO SHORT THAT I had to crane my neck to talk to her. She was even more out of shape than Brett. At first glance, you wouldn't mark Lisa as a cop. At least Brett could have intimidated criminals by threatening to fall on them; what could Lisa do to scare a criminal into compliance? Yet she had been a police officer for almost two decades and had worked the most violent patrol divisions.

Lisa talked softly. Her calm demeanor and ungainliness masked her spirit and courage. She was a fighter. She had joined the LAPD when male police officers were openly hostile toward female officers, when women cops were openly called bitches and were not welcome in the elite police force of macho white males transitioning from a military career to law enforcement. She had passed through that phase of the LAPD and emerged a survivor.

Perhaps to prove herself, she had worked in the roughest neighborhoods of the LAPD's roughest divisions. She had captured the trophy that a hard-charging gunfighter cop most covets—she had stopped a would-be cop killer in a gunfight. A gang member in a crime-ridden South-Central L.A. housing project had ambushed officers, including Lisa and her partner, blazing away at them with an AK-47 assault rifle. With a barrage of bullets whizzing past her, missing her by inches, she had killed the assailant with a single shot from her 9-millimeter Beretta pistol. She had been involved in several shootings, and due to the threats on her life from gang members, she had been transferred to a Valley division far from South L.A.

We sat in our police cruiser. I nervously and unsuccessfully tried to log on to the computer, failing to accomplish what was a mindlessly simple task. My first day on the street in an alien environment had had an inauspicious beginning with Brett's diatribe and then later catcalls by senior officers in the roll-call room as the watch commander ordered me to stand up and introduce myself. I had never been at the receiving end of the sharp wit of street-savvy, foul-mouthed street cops. Their sarcasm was ruthless. Cops love to tease their own!

"Do you know what you are doing?" Lisa asked.

"Yes, ma'am, sure," I replied.

Lisa stared at me intently as I continued getting error messages from the cruiser's computer. Her amused look made me more nervous. Why was I so afraid of her? What could she do to me? Was I afraid of being humiliated by a cop while learning to be a cop? I had always been willing to accept humiliation, suffering, and death as a normal part of human existence. One of the few things I was proud of was my stoic disposition. But my stoicism seemed to have vanished. I was losing my composure, sweat pouring down from my just-out-of-the-academy shaved scalp. Valley heat combined with a half-inch-thick bulletproof vest, a long-sleeved dark blue wool shirt, and thick uniform pants were not helping, either.

"Listen, I don't believe in shouting at my boots or being hard-nosed. You're having a difficult time as it is, and I don't believe in making it worse by insulting you in front of others. Relax, we're partners. You can save my life just as I might save yours. Stop sweating. I am not like other testosterone-crazed male cops. I had to put up with lots of shit when I was a boot. I am not going to do that to you."

"Yes, ma'am. Thank you, ma'am." I did not believe a word she said. Lisa had a tough reputation; boots were fearful of her and gossiped about her being strident. I would have to remain vigilant. I worried that she was waiting to find faults with me and humiliate me in front of the officers.

"Last night officers shot this idiot in front of the 7-Eleven. He is KMA," Lisa said. "You wanna see?" I felt like saying, "Splendid, sure! How did you know that I loved seeing dead bodies?" I knew she was testing me. *You want to see if I am going to throw up on my first day.* Anyway, did I really have a choice? Would Lisa agree if I suggested something more pleasant instead, like going to Starbucks, getting some coffee, and enjoying it in the Japanese Garden near the Sepulveda Dam?

"Sure, ma'am," I said like a true yes-man as Lisa drove to the 7-Eleven store on Van Nuys Boulevard and Vose Street.

"You wouldn't believe this fool. He had been hitting the *same* store for the last five weeks, same day of the week, same time." Lisa sounded incredulous as she stretched *same* each time in her sentence. Even hardened cops could be surprised; it was my turn to be amused.

"Fucker was robbing the same fucking store clerk on Saturdays at eleven P.M. Last night detectives were staking out the store. Guess what, knucklehead paid them a visit exactly at eleven P.M." Lisa chuckled. She was delighted.

"Maybe he was trying to help by being punctual and consistent, ma'am. Maybe he wished to be apprehended." I grinned, trying to be funny to ingratiate myself with this woman.

"Man, what kind of shit do you talk? Where do you come from?" Lisa gave me a hard stare and then burst out laughing.

I was taken aback.

"Weren't you a teacher before you joined?"

I nodded.

"Learn to talk like a cop, kid, or gang members will laugh at you. No one will listen to you. You gotta talk their talk or you get no respect. And if you get no respect, you can't be a cop."

Shit, how do I learn to speak like a cop? Fortunately, Lisa had returned to her story.

"Well, the detectives hiding in the freezer came out. Knucklehead had the clerk at gunpoint. They told him to drop his gun; guess what he did. Motherfucker pointed his gun at the detectives."

I wondered how much the reality TV shows would pay for the store surveillance video capturing the gun battle with a robber getting killed, live on camera.

"Mora shot him once, and the asshole ran out of the store. Actually, he limped out of the store, because he was shot in his ass." Lisa found the incident hilarious; her day, unlike mine, was starting joyfully.

"There were five officers waiting outside; they cornered him right away. Listen to this—the story got even better. This fool was facing five officers. Guess what he did. He looked at the officer with the tube [shotgun] screaming at him to drop his gun, and guess what, he pointed his revolver at this officer. It was a turkey shoot. He won't be robbing anymore, and we won't be wasting money to keep him in jail."

Officers had fired forty-two rounds at the robber, striking him thirty times at close range. Even with the target so close, twelve bullets had missed and were embedded in the asphalt. The robber lay a few feet away from the officers, and when the officers checked his revolver, they found his gun *empty*.

He had been robbing stores with an unloaded gun. But what difference did it make to his terrified victims or to the officers? Having a gun pointed at one's head must be chilling. I was fortunate that I had never faced such an ordeal. At least not yet.

Lisa turned right on Vose Street and parked outside the small strip mall. Yellow crime scene tape blocked the driveway. She lifted the tape and waved me in. A crowd of onlookers huddled near the tape and looked at us. Suddenly I felt a tingling sensation. I felt special; I was doing something I had seen only in movies until this point. I was walking into a crime scene, like a confident Clint Eastwood flashing his badge and walking past a throng of cops to view a dead body.

But unlike the impassive hard-boiled characters that Clint Eastwood plays, I had no idea how I was going to react to this scene. I'd never seen a dead body before, especially one peppered with thirty bullets, sprawled on asphalt for eight hours in a warm Valley night.

Was I going to feel nauseated? Repulsed? Sickened? Was this a display of raw police power, when a single man was shot dead by five officers? What could have been done differently to ensure that no one died? Did the officers rejoice afterward or did they feel remorse? Did the loss of a human life carry any meaning to them? These questions remained, as always, unanswered.

LIKE A DUCKLING, I SILENTLY walked behind Lisa. A Los Angeles county coroner van was parked next to the store. A small group was gathered around a prone body. Lisa was carefully studying my reactions. Any hint of weakness would become a hot topic of discussion among the training officers, and I would never live it down. It would also give fodder to other cops to make fun of me. In my short association with the LAPD during the academy training, I had learned that there is nothing on this earth that cops enjoy more than making fun of other cops. They gladly put their lives in jeopardy for their partners; they also mercilessly teased their partners!

A row of ants was crawling toward the dead body. Flies and other insects were buzzing around the face of the deceased; a strange, pungent smell hung around him. A caravan of ants had entered his nostrils and ear canal and thronged around the holes, creating a shifting mass of darkness over the stiff black skin.

The indignity of death was unsettling. Fortunately, or unfortunately, I was totally numb. It was as if my heart and mind couldn't feel anything. The body of the dead robber was twisted at the waist. He must have rolled while he was falling, and rigor mortis had set in. He was wearing several layers of clothing, typical of gang members and dope addicts. A large pool of blood—he must have lost a gallon—had gelled around his body in a thick layer. The burgundy-purple gel did not even look like blood; it resembled matter oozing from overfilled garbage dumpsters that never get emptied.

Surprising myself, I bent down on my knees to get a closer look. I was not squeamish; instead I was marveling at the fact that a living, breathing human could instantly turn inanimate. Suddenly all of human existence seemed absurd, pointless. I felt sorry for the empty shell lying in front of me; without its spirit, the body had turned into a pot of clay, an empty vessel with no use to the living world. If violent death of *one* individual had such an impact, I wondered how people felt in a war zone. I stared at the shiny pieces of flattened metal stuck in the dirty gray sweatshirt.

"He bled so much that the bullet slugs came out of the holes in his body," Lisa explained. She must have been reading my thoughts.

A smartly dressed athletic woman in her thirties was wiping the chin of the dead man vigorously with a rag to remove caked blood. "I need to take some front shots for his ID," she explained, addressing all of us who had gathered around the corpse.

She was slim and very attractive. Had I met her in a different setting, I never would have guessed that she was a coroner's investigator, with the macabre job of dealing with the dead on a daily basis.

I stepped back to avoid crowding her. She grabbed the chin and top of the dead man's head and violently jerked it. She was trying to straighten the neck. Was she going to break the neck of a corpse? Was it even possible to do that? I'd had enough and turned away. As I glanced sideways, I notice two officers chatting nonchalantly. They were eating McDonald's sandwiches.

Lisa roused me from my state of shock. "Let's go and do some police work."

I thought I had passed my first test. As we drove away, Lisa said, "I want you to know where you are all the time, okay?

"Let's say you get shot and you go on the radio asking for help and you don't know where you are? Not very helpful!"

"Yes, ma'am," I said. I had no idea which street we were on. Policing in 1998 was prehistoric. We had no GPS to help us navigate, no cell phones with Google Maps. One had to carefully monitor the street signs every time the cruiser made a turn and read the block numbers for addresses.

"Also, be alert to the radio *all* the time. It is our lifeline. Answer up immediately when they call us. It is very important to know where you are *and* to listen to the radio. Do you know why?"

"Yes, ma'am, because—"

"They called for us. Did you hear that? Answer the radio. Acknowledge the call they gave us. Tell them we are en route."

"Yes, ma'am." My brown face probably turned red with embarrassment; Lisa was laughing at me. I couldn't decipher whether she was being funny or condescending because I had missed the dispatcher calling for us on the radio.

"Don't worry. It will take you a few days to get used to the radio. Read me the comments of the call."

I looked at the computer screen and read aloud: "Attack just occurred. Female victim eighty-four years old, two suspects, male blacks, late teens, wearing dark hoodies, fled in an unknown direction on bicycles . . ."

Attack was the radio code for rape. Dispatch never said the word *rape* on the radio.

I had been inside a patrol car for only two hours. The first criminal I encountered was a dead black man; the next two were young black males who had raped an eighty-four-year-old woman. I was deeply uncomfortable at the thought.

Was this day an aberration or was it going to be the norm? Had my conditioning begun already? Was I going to turn into a bigot on my first day as a police officer? I wondered as we drove toward the victim's home.

"Keep your eyes open for the suspects," Lisa instructed. "Look for any suspicious-looking black males. They could not have gone far."

What exactly was a suspicious-looking black male? I wondered, but understood that it would be unwise to ask this question right now. We sped to the crime location. Three other police cars were cruising the neighborhood, looking for the suspects. When we reached the victim's house, two police cars were already there.

The day had already turned into the most eventful day of my life.

Molly, the victim, lived in a pleasant residential neighborhood in Sherman Oaks. Immaculately manicured lawns surrounded single-family homes. Tall palms, citrus trees laden with yellow fruit, and magnolia trees with large fragrant flowers graced the yards. Hardy oleanders, whose dark green leathery leaves were overshadowed by sweet fragrant pink and white flowers, hedged the front and rear yards. I had no idea that Los Angeles was such an urban oasis—a botanist's delight.

We sat in a little living room that only a gracious old woman with an artist's aptitude could have decorated. The walls were lined with cabinets full of fine china and collectibles and photographs that were probably more than a hundred years old. Molly sat across from us on a recliner, with her neighbor and old friend, Rose, standing next to her. Molly had lived in this house for fifty-five years. Her husband had passed away twelve years before, and she shared the house with her little dog.

As Molly talked softly, the large bruises on her face and arms troubled me. Who could beat and rape this sweet woman who was old enough to be the heartless criminals' great-grandmother? I sat uncomfortably, not knowing what to say or what to ask for my first criminal investigation. I was relieved that Lisa took over and started interviewing the victim. Lisa listened intently to her story and kept comforting Molly.

"They just broke in. I was working in the yard, I stepped in to get a drink, and they rushed in behind me. I had seen the younger one several times last week. He had been riding on a bicycle. It never crossed my mind that they were watching me. The older one pointed a gun at me and asked for money. The younger one threw me to the ground. He hit me a few times and then he grabbed my breast . . ."

Molly was overcome and couldn't talk anymore. Rose put her hand on Molly's shoulder and Lisa gently touched Molly's arm. I sat stoically, not knowing how to react.

"Is he going to come back and hurt me again?" Molly whispered.

"Don't you worry at all; we are watching your home. We will be watching you twenty-four hours. We are going to catch them. I will drop in again to see you, okay?" Lisa said.

"Thank you so much. With so many officers around, I feel safe now."

What was the point of all this police presence *after* the poor lady had suffered from depravity of the worst kind? I felt a surge of impotent anger. What would I do if I caught this rapist? What created a monster who would assault, rob, and rape a defenseless old woman? What kind of parents raised such children? Did this boy gain pleasure or satisfaction by brutalizing an utterly helpless person? Was he out bragging about his exploits to his friends? What kind of friends would shield such savages? How could this scum's partner watch while Molly was being raped? Would I care if he was executed?

I had been told numerous times not to take anything personally in police work. I needed to focus on what was in front of me. I couldn't get distracted. Distraction could get you hurt or killed in police work, or worse, cause you to lose control over yourself, lashing out at a suspect. But it was hard not to take some things personally.

The ambulance had arrived to take Molly to the hospital. Lisa patted Molly's arm. "We will be watching you. We're going to find these animals."

I stood up and gave a faint smile to Molly and Rose. We were not the primary unit for the rape call; Lisa had helped them by interviewing the victim, so our job was done. After the rape incident, we got a Code 3 domestic violence call.*

"BE CAREFUL. I HAVE BEEN to this place before. This knucklehead beat up his girlfriend a month ago and was arrested. He is an asshole. They obviously got back together. These women get the shit beaten

---

* A Code 3 call is the highest priority emergency call where an officer is allowed to ignore the rules of the road while using lights and siren.

out of them and then go back to the same men. I just wonder!" Lisa shook her head. "Remember, DV [domestic violence] calls are the worst kind. You have to take charge instantly. Sometimes when we hook up the boyfriend, the girlfriend jumps on us and the fight is on. It is never safe at DVs. Stay out of the kitchen and keep them away from the kitchen. Too many weapons in the kitchen. Make sure of where you are. Maintain command presence. Remember, you have to run the show. If you lose control, you will escalate the situation and start a fight. Pull the man away from the girlfriend right away; keep them apart. Hook him up right away before we do anything. Remember your tactics. Okay? Understand? Are you ready to rock and roll?" Lisa stared at me with an intensity I hadn't seen her display yet.

Lisa had basically read me the entire manual on domestic violence while speeding toward the location.

A neighbor had called 911 to report a screaming woman, possibly being assaulted by her boyfriend. Lisa parked half a block away from the apartment complex and we ran to the location. Three little children were crying hysterically outside the door.

Lisa pushed the door open, startling the couple standing in the living room. Staring at us in surprise was a tall, muscular Latino man with a shaved head and gang tattoos. He was wearing the clothing of a gang member: a white vest and sagging oversize pants held up with a thin belt

I was profiling a man without knowing him, just by his clothing and appearance. This was not even the middle of my shift! I couldn't think like this; I needed to focus. The woman darted from the living room to the bedroom, disappearing from our view.

"Turn around and put your hands behind your head," Lisa screamed.

"Why, I haven't done anything. Isn't it true, honey?" The man turned toward the bedroom and shouted, "I didn't do anything. Isn't that true, sweetheart?"

The man was ignoring Lisa; he was not complying with her commands. He had placed both hands on his belt, and he had turned and looked toward the bedroom. I didn't know what to do. Should I run inside and grab his arms? But he was much larger and stronger than I was. Should I take my pepper spray out before I approach him?

"Turn around and put your hands behind your head or you are fucking dead." Lisa had her gun out. "Get down on your knees, you asshole. Right now! Keep your fucking hands behind your head. Don't move or I'll cap your ass," Lisa screamed.

"Don't shoot, don't shoot, I am going down. I didn't do nothing, ask her." The man continued to protest as he went down on his knees.

"Shut the fuck up and stop threatening your girlfriend. Don't say a word to her." Lisa looked at me and said, "Go and hook him up."

I walked tentatively toward the man. I fumbled while removing handcuffs from my duty belt. I was handcuffing a suspect for the first time in my life. The man by now had started crying. "I didn't do nothing! Ask her, ask her—sweetheart, isn't it true?" It was all a show; he was instructing his girlfriend what to tell us.

"Put your hands over your head," I said, trying to steady my voice.

He did, adding again, "Ask her. Sweetheart, isn't it true?"

I pulled his arms behind his head and hooked him up.

"Take him out. He is trying to intimidate her," Lisa instructed.

I helped the suspect up and walked him out of the apartment.

Our backup had arrived. They transported the wife-beater to jail while we took a report from the victim. I started the preliminary investigation report and began asking questions.

The poor woman was distraught. "He didn't do anything. He didn't mean to hurt me. Please let him go. I am not going to press charges."

Lisa stared at me impassively as I completed my investigation and offered the victim a protective order. She didn't speak a word while I talked to the victim. It was only when we walked to our cruiser that I became the object of her fury. Lisa gave me a dressing down in our cruiser.

"What the hell were you waiting for at the apartment door? He could have gone after his girlfriend; he could have taken her hostage. He could have killed her in the bedroom with two LAPD fuckups standing out there like idiots. What the fuck were you thinking? Why didn't you control the situation? Didn't I tell you to take control before we arrived here?"

I stared down in humiliation.

I had stood cluelessly while watching Lisa order the wife-beater to his knees. I had no excuse. I watched the entire incident without taking any initiative to control the situation. Things could have gone south in an instant because of my mental paralysis.

"Don't fret over it." Lisa finally calmed down. "What you need to do is think *while* we are driving to an incident. Think in advance about how you will react when shit goes sideways. Once bullets start flying, it is too late to think about tactics. *Think* tactics all the time; think tactics when you are driving to work, think tactics when you are watching TV, think tactics while you're sleeping. Run scenarios through your mind. You don't get a chance to fix things in this business after you fuck up. You will get someone killed or get your partner hurt if you have your head up your ass. This is not a classroom where kids listen to you. You have to *make* knuckleheads listen to you. Got it?"

"Yes, ma'am." I was melting in shame; I couldn't even look in her direction.

Why wasn't the earth opening to swallow me? First day on the job, and I had been thoroughly disgraced. But there was no time to wallow in embarrassment; we had to book the wife-beater and complete the arrest report.

We returned to the station to finish booking the suspect. After the booking, we were en route to our next call, where it was now my turn to hold the hands of a bloodied victim and comfort him.

My academy instructors had warned me that Van Nuys was a busy division. It indeed was.

———————

THE DAY HAD BEEN INTENSE. Between the wild crime stories, the dead body of a robber, an octogenarian rape victim, unknown streets and alleys, unceasing incomprehensible chatter on the police radio, and hazing by senior officers, I had been exposed to a universe that I never knew existed—a universe that was challenging me to open my eyes and examine my deeply held beliefs. On my first day, I had been too busy trying to learn and survive on the streets of Los Angeles. During the eight-hour shift, I had not learned anything to affirm or contradict whether cops were racist brutes who treated minorities with contempt or brutalized inner-city youth. Lisa certainly did not fit the stereotype. Yes, cops were callous and aggressive—I guess one couldn't survive daily exposure to violence or danger without these traits. What, then, separated aggressiveness from brutality? Police work was considerably different from what I had imagined or had been told about. Lisa's hand on Molly's arm was the evidence that one *could* do "good" as a police officer. Maybe I shouldn't have been surprised at this observation, but I had come from a vantage point that saw cops and authority as nothing but oppressive forces. I had also not learned anything that would help me understand Raju's behavior.

The robber whose body I viewed on my first day in the LAPD was a poor young black man who lived in a high-density low-income apartment complex not far from where he met his end. Could I blame poverty and lack of opportunity for his descent into a life of crime? If so, then how would I explain the choice of his own neighbors, the blacks and Latinos who lived in the same underprivileged circumstances but refused to pick up a gun and commit robberies? Did the rapist have an abusive childhood or was he a victim of poor parenting? Even if he did, why should it excuse his behavior? Or that of the Latino gang member who was beating his girlfriend? What could I learn from them that could help me understand Raju's behavior? Was the lesson as simple as that some people are different? That they are genetically predisposed toward a life of crime and antisocial behavior? Or did their families raise them to be criminals? Or did their cohort, the peer group, and their surroundings influence their behavior? Or a combination of factors? What could have been done so that the robber never picked up a gun in the first place, preventing his own violent death? What could the rapist's parents have done so that he did not violate a defenseless elderly woman? Could my parents have done anything to stop Raju's slide into a life of delinquency? Was there anything that I could have done? I had looked for the answers in science, psychology, philosophy, and sociology; now I was searching inside the criminal justice system. No one seemed to have the answer. I needed to dig deeper for clues.

On the way home, I decide to follow J. Krishnamurti's advice and be an observer, at least for one more day. Despite all my fears and insecurities, I knew that I would show up for work the next day. I was willing to pretend to be a police officer, at least once more.

12

VICTIMS AS KILLERS

Surprising everyone, especially myself, I had somehow completed two months as a recruit. I was now assigned to my second training officer, Andy. Andy was very different from Lisa. She carried herself calmly; Andy was hyper, even childish at times. Lisa showed compassion toward suspects and gang members; Andy did not tolerate criminals. Lisa spent the shift taking care of the radio calls and rarely engaged in observational policing; Andy's mission was to go out every day looking for the bad guys—and finding them. With twenty years under his belt fighting crime and surviving on L.A.'s rough streets, Andy was a true street cop. Riding with Andy guaranteed footage of wild police work, kicking down doors, chases on foot of men with guns, wild pursuits of criminals, knock-down, drag-out fights. The television crew for the reality show *LAPD: Life on the Beat* loved to ride with Andy; he was their man. The crew loved to shadow me, too, although for a different reason. I was the opposite of Andy, but still a good story: the fumbling raw recruit lacking street smarts. I didn't even talk like an average normal person. I had difficulty conversing with gang members and suspects. When people listened to my academic

English, they looked at me as if I came from a different planet. I had a serious problem exercising authority. On my first day with Andy as my training officer, and with the crew holding a camera and a boom mike near my face, I must have looked absurd interrogating a gang member.

"Sir, what is your last name? Why are you standing at this intersection? Do you belong to a street gang?" I asked a dressed-down gang member hanging around on Blythe Street, where there was a court-sanctioned injunction against the congregating of gang members. The gang member burst out laughing. He was barely fifteen. A large brown-plaid flannel shirt floated over his thin body. His dark blue pants were too large for his frame. They could be hiding a pair of shotguns. Six inches of his white cotton boxers were revealed as his baggy pants continued to slide down. He kept pulling them up every few seconds. He looked eerie with his shiny head and tattoo-covered neck and scalp.

"You rookie, man?" asked the kid, giving me a condescending smirk. The little gang member had correctly guessed that I was fresh out of the academy, but probably anybody could have seen that I completely lacked street smarts. The TV crew tried to suppress their laughter. I was dumbfounded and humiliated.

"Why you harassin' me, man? I ain't doin nuttin, man," he added.

"Hold on. Let me deal with him." Andy moved in; he could see the gangster was not only humiliating me but was also getting away with it. "What do they call you? Who you kicking with? Don't bullshit the bullshitter."

While Andy went after the banger, I wondered what Andy was thinking of me as a potential LAPD cop.

The fact that the camera was recording my ineptness intensified my embarrassment. My week under the lens, with the film whirring and a long boom microphone thrust at my face at each traffic stop—or indeed, any time I opened my mouth—made me feel as if I was under a microscope. It was hard enough to inter-rogate and assess, let alone pat down and handcuff, street-smart, slang-speaking gang members without having to worry about my performance being evaluated by a hard-charging training officer and a battle-hardened TV crew that had seen it all.

I SAT WITH MY FELLOW officers while the sergeant called roll and as-signed us our cars. I had been a rookie now for two months. Andy and I worked in an area that covered Blythe and Valerio Streets. Besides Langdon Street, Blythe and Valerio were known for gang violence, drug-related problems, and violent crime. Cops working these streets stayed busy all night chasing the never-ending radio calls about drug deals in progress, drive-by shootings, assaults with a deadly weapon, and gang activity.

After the roll call, Andy drove straight to Yoshinoya Beef Bowl, his first stop before starting the shift and taking radio calls. While he was good-natured and a pleasure to work with, he was cranky when he was hungry. The teenaged Vietnamese waitress knew him well. As we started to eat, I heard a broadcast over the radio:

"Nine Adam twenty-three, 211 in progress, Van Nuys and Valerio, suspects several male Hispanics, dressed in baggy clothes, 9-A-23, respond Code 3."

Lunch was over. We were 9-A-23 and were being called to the intersection of Van Nuys Boulevard and Valerio Street where a robbery, a 211, was occurring. "Respond Code 3" meant "drive at lightspeed with red light and sirens on, dodging the reckless L.A. drivers." Officers loved Code 3 driving, with its adrenaline rush *and* legal exemption from all rules of the road. I considered it the equivalent of a close dance with death.

We left our food on the table, ran to our cruiser, and jumped in. Our cruiser rocketed out of the parking lot, siren blaring. Driving at lightspeed required fast thinking, since most Los Angeles drivers seemed to lose their common sense whenever they stepped in their cars. I frantically threw everything on the floor—ticket books, logs, clipboards. Next I strapped my seat belt on and did the same for Andy. His job was to drive to the location as fast as possible, be ready with a tactical plan when we arrived at the crime in progress, and avoid accidents on the way. My job was to navigate, broadcast on the radio, and alert my partner to traffic.

"Nine Adam twenty-three, responding Code 3 from Van Nuys and Haynes, requesting backup and an airship (the LAPD helicopter)," I radio in. We always requested a helicopter for serious calls, in case suspects took off in their car, starting a car chase. A helicopter, with its aerial 360-degree vantage point, almost always guaranteed the capture of suspects on the run or hiding from pursuing officers.

"Okay, Andy, right is clear, keep going, keep going, WATCH IT, watch it, watch it, slow down, slow down, intersection coming, traffic right, traffic right, okay, clear, keep going." During an emergency run, we worked as one. Andy relied on my judgment and I on his.

Our cruiser was speeding on the six-lane-wide Van Nuys Boulevard, which was clogged with heavy daytime traffic. Andy sped down the median, siren wailing, red and blue strobe lights whirling, causing the vehicles ahead to open up and make way.

"Andy, A-59 and 11 are backing us," I updated my partner. "Airship has not responded yet."

The police helicopter had not acknowledged our request yet, but two backup units were on the way. Andy was driving like a maniac. He loved it. I obviously hadn't learned to be a cop yet, because I didn't find the death-defying near misses fun.

Just before we reached the intersection of Van Nuys and Valerio Street, Andy turned off the siren. Unlike what you see in Hollywood movies, real-world cops don't drive at Mach speed with sirens blasting directly into a crime scene. Suspects could determine a police car's exact location by the siren and lights, making cops an easy target.

I saw a crowd on the northwest corner of the intersection by a small strip mall. Several small stores lined the strip, including the friendly neighborhood bank with special twenty-four-hour service for small-time robbers seeking quick withdrawals—in other words, a 7-Eleven.

Andy made a quick left turn and parked south of the mall on Valerio, taking cover behind a wall. Before our car came to a halt, I grabbed the shotgun from its rack, kicked the door open, and jumped out. Andy jumped out the other side of the cruiser. This was not done to impress onlookers. The worst scenario of all was to be ambushed while you were trapped in a car.

Our backup units pulled up behind us as we inspected our surroundings. There was no sign of any crime in progress. Since I held the shotgun, I led the group. The others followed, their pistols drawn. We all held our guns in low ready position—fingers on the slide, ready to move to the trigger, barrels pointing toward the ground until a target showed up.

We cleared the parking lot from the south and moved slowly toward the 7-Eleven. My trigger finger was on the shotgun safety. My heart was beating fast, and a nervous energy radiated throughout my body. I could see that the front glass door on the 7-Eleven was shattered. Fragments of glass were scattered across the parking lot. A red four-door Honda Civic parked in front of the store looked as if a herd of elephants had trampled it on their way to buy Slurpees. The roof was caved in, the hood and trunk were banged up, all the windows were broken, and the car was freshly scarred with the graffiti of the Valerio Street Gang.

I didn't see any bullet holes or bodies sprawled on the ground; there was no fresh pool of blood on the pavement. A colorful mélange of residents and gawkers had separated into several clusters at the edge of the parking lot. They spoke in various languages, shook their heads, looked concerned, numb, fascinated, grim; some were laughing nervously. The onlookers represented many backgrounds—Caucasians, blacks, Vietnamese, Filipinos, Latinos, Indians, Koreans, and Chinese. They all lived or worked here, where violence erupted periodically, often at the slightest pretext.

"Get back, get back," the cop behind me bellowed. "Anybody hurt? Who saw the robbery?" We moved to corral the bystanders, trying to distinguish between victims and witnesses in the group. Perhaps the perpetrators were trying to blend in. Surprises crop up from the unlikeliest of the places, and being unwary or careless could be lethal.

I pointed the shotgun at the door of the 7-Eleven. "This is LAPD. Everyone inside, come out one at a time with your hands up!" I shouted. Robbers could be hiding inside or holding hostages. Jasbir Singh, the clerk, slowly walked out, his palms clutching the top of his black turban. His eyes went wide and his knees wobbly as he looked at the shotgun barrel pointed at his torso; he started shaking, almost falling to the ground in fear.

Jasbir was a Sikh with a flowing black beard and a black turban concealing his three-foot-long braided hair. "Step out slowly with your hands up, turn around, and walk back slowly toward me."

"Don't shoot, don't shoot!" he screamed. "They are gone, they are gone." He didn't want to be shot by a nervous officer.

Five officers with guns drawn made a rapid entry into the 7-Eleven. The store must be checked, as suspects might be hiding or victims might be lying dead inside. For cops to trust even the victims and let their guard down could become an invitation to suicide.

But no one was inside the store.

With neither suspects nor victims in sight, we spread out and started scoping for witnesses.

"Is there anyone hurt?" I asked. "Who saw this? What happened? Sir, if you have no business here, please move."

A blond middle-aged white woman walked over to me. She was the manager of the auto parts shop next door. She spoke excitedly in barely comprehensible sentences, gesticulating wildly.

"These six or seven gang members, they were hanging around right over there. Two of them came inside and bought paint cans, I am sure for tagging. They were shaved-head gang members—you know, white undershirts and baggy jeans. The whole group was hanging out near these pay phones. I knew they were looking for trouble. I was keeping an eye on them. Then this red Honda pulled in and two men got out. They exchanged words with the group. Gang members went after them. I thought they were going to kill the two. Both ran into the 7-Eleven and locked the door. Two or three gang members were trying to break open the door, and the others were smashing the Honda. They all ran when they heard the police sirens. Had you not come here so quickly, people would be lying dead. I'm so relieved that nobody was killed."

"Ma'am, did you see any weapons? Any guns or knives? What happened to the Honda driver and his friend? Where are they?"

"I didn't see any weapons; I wasn't that close. Ask the 7-Eleven clerk. He was right in the middle of it."

From the eyewitness accounts, it was obvious that some Valerio Street gang members attacked two men, who barricaded themselves in the 7-Eleven to save their lives. We had no good suspect description. The descriptions from the statements—"shaved heads, male Hispanics, baggy pants, white T-shirts or undershirts, sixteen to twenty-one in age"—fit a multitude of people in the area. We didn't know who owned the red Honda. The car was registered in Pacoima and was not reported stolen.

While Andy and the other officers talked to other witnesses, I walked with Jasbir into the 7-Eleven. This twenty-four-hour store was a favorite coffee stop for cops, especially the graveyard shift. I knew most of the clerks here. They were from the backwaters of Punjab, most of them farmers who left their land, their families, and their way of life to come to the United States. These uneducated and unskilled peasants, with English skills rudimentary at best, were enticed by the stories of opportunity and easy money. Some worked twelve-hour shifts seven days a week for a minimum wage, occasionally getting beaten or shot by criminals; but they took pride in saving money, which they regularly sent to their families in India.

Most of them did not trust America's judicial system, but that was not the fault of the LAPD. These Punjabis had known only corrupt Indian police and thus were wary of police officers.

Jasbir Singh grabbed a broom and started cleaning up the shattered glass. I spoke to him in Punjabi, to make him feel comfortable and win his cooperation. Like most eyewitnesses to gang violence, he did not want to get involved. He worked in a store frequented by gang members. He was aware of the execution-style murders by Asian Boyz gang members that took place in the apartment complex right across the street from his store. He also knew that the police arrived *after* the crime, when the victims had been stuffed in a body bag or carried by an ambulance to a trauma center and the criminals had disappeared. He had to survive on his own when we were not around, which was most of the time.

"Hey, Jasbir, tell me what happened. I need to know, man. What did you see?" I asked him.

Jasbir was sweeping the glass shards and answered evasively. "These two Mexican boys ran inside. One of them grabbed the door handle. Gang members outside smashed my glass door. Other Mexicans were banging on that red car."

Los Angeles is rich with diverse Latino populations. People from Guatemala, El Salvador, Nicaragua, Colombia, Honduras, and several Central and South American nations live in the city; they hate to be lumped together as "Mexicans." But to many Angelenos, every Latino is a "Mexican."

"Let's look at your surveillance video. We may know who these troublemakers are. They need to go to jail." I tried to encourage him, but I knew it was a losing battle.

Jasbir solemnly looked at his broom. "Sunil-ji"—*ji* added to my name as an honorific—"please wait for the manager. He will be angry with me. The video recorder is locked in the storeroom. I don't have the key."

I knew I wouldn't get anywhere and gave up. The video may have all the parts of the puzzle, enough to jail these gang members. Jasbir promised to save the video for the detectives. I felt disappointed that people didn't cooperate with the police, letting the criminals get away. But people living in crime-ridden areas had to coexist with the gang members, drug pushers, and bullies. We could not promise protection from the criminals. Witnesses to murder, shootings, arson, and assorted crimes often refused to show up in court, and if they did, they changed their testimony to protect the criminals. It is disgraceful that hardened criminals successfully intimidate witnesses and the police can't overcome people's fear. We fail to protect or serve the most helpless members of our communities.

Unable to find a victim, we couldn't make a report of assault or attempted robbery. No victim, no crime. Our backup units left. Andy went into the 7-Eleven to scratch out a quick vandalism report on the shattered glass door. I took out a vehicle investigation form from my folder. With its smashed windows and bits of glass, this car was a hazard. It needed to be towed to the police pound.

Andy walked out of the store after getting Jasbir to promise (again) to save the store surveillance video. The form I was filling out was almost done when I observed a short Hispanic male walking toward the Honda. Very athletic-looking, he was about five three and around 140 pounds. He had a military style crew cut and wore a body-hugging white vest and form-fitting blue jeans.

"Stop. Step back from that car," I said to him. "What do you want?"

"*No comprende.*"

Half of the population served by the LAPD speaks only Spanish. It took me less than a month to forget my seven months of Spanish education in the police academy. In fact, the only "academy Spanish" I had retained were bare-minimum survival commands, used to take suspects in custody. I could say very fluently "*Suelta el arma, manos arriba, despacio voltéese. Ponga sus manos atras su cabeza. No se mueva.*" Drop the gun, hands up, slowly turn around. Put your hands over your heads. Don't move. But I was incapable of carrying out a conversation in Spanish. Luckily, Sergeant López was on the premises talking to witnesses, so he took over.

"His name is Alfredo," he briefed me. "It's his Honda. Four or five Hispanic males tried to carjack him. He was here with his girl-friend. They both ran inside the 7-Eleven, and he held the door handle to prevent the suspects from getting in. The others trashed his car." The Valerio Street Gang graffiti on Alfredo's car had pretty much already told us who the suspects were.

Had we been a few minutes later—stopped to wrap up our lunch, for example—we would have walked into a homicide.

I had a strange feeling about our victim; something didn't feel right. Why did he disappear when the police came? Why did he return almost thirty minutes after our arrival? Why did the suspects destroy his car and carve gang signs on it? Where was his girlfriend? No one mentioned a female, including Jasbir, who had no reason to lie to us.

Despite our doubts, he was still our victim and we took a crime report. Alfredo would have to get his own tow. Two hours after our interrupted lunch, we returned to the station to finish the paper work. In the grimy, gray break room, Andy plunked some coins into the junk food machine, which spits out a brownish hot dog. It's seven P.M. when we leave the station and resume patrol.

———————

AS WE DROVE ON THE busy streets of Van Nuys, my mind wandered. I remembered a conversation I had with my friend Lynne when I was a doctoral candidate at the University of California campus at Davis. The discussion had somehow turned to inner-city crime. I was emphatic that racism and exploitation by the white power structure had created the gang problems. Lynne refused to give any weight to my arguments. She insisted that gang members were petty criminals destroying their own neighborhoods. I liked Lynne a lot for her progressive views about society; however, I could not understand how she could be so callous about the poor misunderstood inhabitants of the inner city. Perhaps she was reflecting the prejudices of her privileged upbringing. How could a young white Northern California girl know anything about the inner city anyway?

Four years later, after leaving Davis, I saw Lynne's face in Danisha, my assistant for a biology class I taught in a Los Angeles junior college, located in the seediest part of the town in South Los Angeles. My class was incredibly diverse—the only ethnic group missing was Caucasian. Between the breaks, Danisha and I talked about many things, including her boyfriend, who had already racked up a decent police record.

Danisha's opinion about crime in the inner city had surprised me. This inner-city black teenager hated gang members, calling them losers and cowards. How could I discount her views as I had Lynne's? Danisha lived in gang- and crime-ridden South-Central L.A.! Her philosophy about crime was the opposite of mine. I used to think that gang members were just misunderstood kids asking for attention. She vehemently disagreed. "They ain't good for nothing.

You think they be nice to you if you nice to them? Ha, they gonna chew you up and spit you out. You don't know these fools. What they need is some serious ass whuppin'. That's all they understand. They ain't no rebels, they cowards—ain't never step away from their corners, ain't never step out from the hood."

Despite growing up in poverty in India, near slums, I never had to face crime's ugliest face. My upbringing had been sheltered. As I moved through graduate school in Florida and then California, I had perhaps become a naive intellectual. In my opinion, the dread-locked gangsters in South-Central Los Angeles were rebelling against an occupying army of police and the economic injustice of a capitalistic society. Latino youth were up in arms against racism and police brutality. If we showed them compassion, rather than crushing them under the heavy hand of the law, they would change. Good jobs would keep them away from a life of crime. But there is much more to the problem of gangs and violent crime than what liberals like me or reactionary conservatives think.

As I came to know gang members, I learned that they had no knowledge or understanding of gang culture or history and abso-lutely no interest in improving their own lives or the communities they lived in. They were busy destroying their neighborhoods with violence and drugs. When not bullying people, they were driving businesses out of the neighborhood by extortion. I tried to reason with young men whose horizon did not extend beyond guns, sex, low-rider cars, alcohol, and partying. They bragged about the "re-spect" they commanded, compared to the gang in the next block. They had large egos and were obsessed with watching their own exploits on the evening television news. These young men craved media coverage. I met kids scarcely fourteen years old who had shot someone to earn *respect*. Demanding respect via the barrel of a gun? Exercising power through fear, intimidation, and violence? What kind of rebellion against society was this?

These youngsters did not understand or care that they were killing their own people and destroying their own communities. The Latino and black communities have suffered racism and invidious discrimination, but gang members exacerbated these social problems; worse, they pretended to be victims of the "man," as if the dominant social group had forced them to kill the kids in the next neighborhood. As I learned more about them, I came to see the gang members as criminals, financing their mayhem and criminal lifestyle with drug pushing and extortion. It was a remarkable about-face in my outlook.

Whenever I walked through housing projects, the mythical "hood" where poverty, broken families, and social neglect supposedly caused youngsters to band together in a violent brotherhood, retaliating against the smallest of slights, I marveled at the wealth that existed there. Billions of people on this planet would instantly switch places with ghetto residents, ecstatic to move into apartments that had color televisions, refrigerators, dishwashers, running water, twenty-four-hour electricity, and functioning toilets. The poorest of the poor in Los Angeles housing projects lived like royalty compared to the poor of the world.

I had spent my entire childhood without *any* of these amenities. I grew up in a house where the water taps ran barely for two hours a day—and we were more privileged than the people in the slum near us; they had to line up to get water from a single communal tap. Over 700 million people in India still don't possess what our ghetto residents have, yet they don't go around shooting, stabbing, selling dope, or creating mayhem. About 200 million untouchables and other downtrodden caste members in India, the poorest and the most vulnerable there, have been oppressed for centuries. The treatment of these Dalits, or "broken people," by the upper castes in India is no different from the treatment of the slaves by Southern slave owners in the antebellum South. Yet the untouchables don't go around destroying their own communities. They don't rob or kill their own community members and make their communities even more miserable.

Poverty, abuse, and oppression do not legitimize reckless violence and destructive behavior! It is immoral to make any excuses for such violence and self-destructive behavior.

It was embarrassing to reflect back on my assumptions. Two years ago, when I knew nothing about crime and the inner city, I thought I had all the answers. Now I was in the thick of it; I had no idea what was going on. What I did realize was that change sought through peaceful Gandhian nonviolent ways was the only workable solution to fix our violent society. Gang violence, retaliation, and further retaliation only spill blood and keep creating more victims.

—————

ANDY AND I DROVE AROUND for the next four hours, dealing with family disputes and burglar alarms, and keeping an eye on Blythe Street gang members. It was eleven P.M. and our shift was nearly over. Andy turned onto Van Nuys Boulevard and headed back to the station. I began closing my log for the day. Another day was over, and I was going home safe. No one shot at me, and I didn't get involved in a knock-down, drag-out fight with a criminal. This makes for a great day in a cop's life. Evening traffic had thinned down considerably, with few stragglers left on the street. I had started unwinding. Hours of remaining alert are hard on one's nerves. So much adrenaline pumped through my veins during the shift that it was hard for me to fall asleep for hours after the shift ended.

As we drove past Saticoy Street, Andy pointed left to the southeast intersection of Valerio and Van Nuys. "Look at those five shitheads," he said. "Keep an eye on them, man. They're up to something."

I saw five dressed-down gang members splitting into two

groups. Three were crossing the intersection, walking toward the liquor store on the southwest corner. The other two had blended into the dark near the pay phones on Van Nuys Boulevard, just a few feet south of Valerio Street. Instead of stopping to check what was afoot, Andy kept driving.

"Look straight ahead, man, I'm going to pretend I didn't see them. We'll sneak up on them."

To the uninitiated and the ignorant, Andy might have been indulging in racial profiling. Oversize clothing, shaved heads, and tattooed Latino teenagers are not necessarily gang members. However, his suspicion was based not only on their appearance but also on the location (an area with a high crime rate and a lot of gang activity), on the time of day (late in the evening, when most violent crime occurs), on the context (a gang incident earlier in the day at this intersection), and on the behavior of these five young men (they had split into two groups, with one group acting as a lookout). In fact, Andy was engaging in proactive policing, which society desperately needs: preventing crime by using his knowledge, skills, and experience of gang violence.

Andy drove over the curb and parked on the sidewalk south of Valerio Street. Darkness gave us good cover. The three gang members, now in the liquor store, couldn't see us. We stealthily approached down Van Nuys Boulevard, staying close to the walls and closed storefronts. The liquor store was in a small strip mall. On its north was a Salvadoran restaurant, now closed for the night. Several small stores bordered it on the east. Its storefront floor-to-roof glass windows displayed large posters showing seductive, barely clad young women extolling the virtues of various alcohol brands.

We peered into the store and saw the party of three wandering aimlessly, looking at liquor bottles. The only other customer in the store was paying at the cash register.

"Sunil, stand to the side here and watch my back," Andy says. "I'm going to check them out."

I took cover behind a large concrete pillar from which I could watch the store and the people inside. Andy went in, walking straight toward the kid who had made him suspicious in the first place. The cashier, a Middle Eastern man, was helping the customer. Things seemed fine. I was hoping that Andy would check out these young-sters and we could go home. It was close to the end of the shift, I was almost growing complacent. I didn't even consider calling for backup; in fact, I didn't even radio our location in case of danger.

Already I had made several mistakes. Let me add another one: I was concentrating on what was happening inside the store, watch-ing my partner's back, but not scanning the surrounding area for the lookouts.

Andy did a quick pat-down search for weapons on the first two men and sent them outside. He started talking with number three, a short, stocky Hispanic male. Number one and two came out and walked directly toward me. Number one was a tall, skinny Hispanic kid, five ten, 120 pounds, eighteen years old. His head was shaved and his face deeply pitted by severe acne scars. A dark brown-and-black-plaid shirt, ten sizes too big, hung on his frame. Beneath his shirt, he wore the familiar trademark of the Latino gang member—the white undershirt. His oversize black pants were held up by a thin leather belt. His pants cuffs were worn from being dragged along the ground when he walked.

His similarly dressed companion was short. His narrow cold eyes, crooked nose, and slightly disfigured jaw made him distinct. One quick glance at his face and I knew I would never forget it. A sixth sense was warning me as they sauntered toward me. I assumed they were unarmed, since Andy had patted them both down. Their eyes were latched on me as they walk past and stopped three feet away from me. I needed to maintain a position of advantage so that I controlled the situation and had time to consider in advance how to respond to abrupt and unanticipated actions. Academy instruc-tors had drilled that into our heads: "Anyone standing closer to you than the distance of your extended arm with baton pointed outward has violated your safety zone."

They were well inside my safety zone, close enough to grab the gun from my holster. Their proximity left me almost no time to react. An attack would have been impossible to beat off. I realized they had taken a strategic position, forcing me to turn away from the store. My back was now to my partner, whose back I was supposed to be watching. But I had no choice. I must respond to this threat beside me first. I tried to take command of the situation.

"Step back. Step away from me."

Number one and two ignored my command. Instead, they began firing questions at me at a rapid pace. "Officer, is something wrong?" "What's happening?" "What's the problem?" "What's going on, everything okay? What are you doing here?"

Their rapid-fire questions were a distraction technique. They couldn't care less about the answers. All they wanted was to keep my eyes off my partner *and* their partner. Instinct told me an attack was coming any moment. My brain had unleashed a cascade of adrenaline in my bloodstream, my heart was racing, and my bulletproof vest was soaked with sweat. In a fraction of a microsecond, I needed to accomplish many things. I must control the situation, get this duo away from me, check on my partner—who mistakenly believed that his back was being protected by me—and radio our location in case we couldn't handle this situation by ourselves.

"Right *now* step away from me," I say. "I am not going to say it again." I tried to sound loud and convincing. As I spoke, I stepped into a side stance and grabbed my collapsible baton, ready to swing it open and strike if they advanced toward me.

From this position, I could keep my eyes on them while using my peripheral vision to look inside the store. They backed off a little. I glanced at the window. Number three was facing Andy while walking backward, his outstretched hand holding on to something. Like these two in front of me, the third man was also "dressed down" in classic oversize gang-member apparel. I couldn't see his face.

There was something wrong. Andy was walking toward him, saying something. It looked like number three was trying to get away from Andy, who might not have finished searching him. My gut told me to get into the store.

The two standing near me got even louder with their two-man-con duet. "Officer, what's going on?" "Look at us, man." "Hey, talk to us." "What's wrong, Officer? Look at me, here, tell me!"

Number three ran out of the store and headed north on Van Nuys Boulevard. I took off after him. Andy came running out of the store.

"Watch it, he has a gun!" Andy shouted. I was in a foot pursuit, chasing an armed, fleet-footed small bundle of muscles wrapped in oversize clothes. He was fast.

I heard Andy coming after me, shouting into his radio on the run: "Nine Adam twenty-three, we are in foot pursuit of a man with a gun. We are northbound Van Nuys, requesting backup and an airship."

Number one and two had apparently joined in the late-evening all-out sprint as cheerleaders. I heard them screaming: "Run." "Run." "We'll get it." "We'll get the gun." "Run."

I picked up speed, trying to keep up with the fleeing armed gang member. I was not even thinking about his gun, even though he could have easily turned around and shot me. My gun was in my holster; I was at a disadvantage. If Mr. Fast turned to fire, my brain would need seconds to process this crucial piece of information; then it would need to issue a command to draw my pistol, sight and aim at the suspect so as not to fire wildly at some building fifty feet away, and then squeeze the trigger. These precious seconds would give Mr. Fast time to kill me and make a getaway.

Yet I had decided not to run with gun in hand, leaving it holstered. I'd rather not risk tripping and falling during a pursuit, which could result in an accidental discharge striking a bystander. Both options were fraught with risk. I had made my choice; there was no time to second-guess myself or be analytical.

I saw him turning left on Valerio Street. I was about thirty feet behind him. I couldn't hear the distinct sound of heavy boots striking the ground or the clanging of the baton against Andy's duty belt, which meant he was not directly behind me. But I could hear Andy's voice on the radio; he was doing exactly what he should. In a foot chase, the officer in front chased the suspect and the partner behind radioed for help and gave directions to the backup units rushing to the scene.

I had been running maybe five or ten seconds, but in that time warp, a million things had gone through my mind. Could I catch number three, who was extremely athletic? Was he going to ambush me around the corner? Were the two other gang members still behind us? What about the two lookouts? Did they have weapons? Were they coming after us?

At this point, I had been out of the academy only three months. I was thirty-four, in good shape, and a fast sprinter. In fact, I used to be *proud* of my sprinting ability—until I learned that a suspect on the run had a much greater incentive than a police officer did. The prospect of avoiding jail puts wings on one's feet. Another painful discovery was that heavy police boots and thirty pounds of gear strapped to the waistband of a Sam Browne duty belt would slow down even an Olympian. I carried the essentials every cop must have on him: a belt with a fully loaded pistol in its holster; ammunition pouches with two extra magazines; two pairs of handcuffs; one collapsible baton; pepper spray; a police radio; a small tape recorder; a large flashlight in the rear pants pocket; and a concealed five-shot backup revolver in the front left pants pocket. I had not trained for chases equipped like that!

I turned the corner . . . relief, no ambush. I was now chasing him heading west on Valerio. The street was dark. The large flashlight fell out of my pants pocket. The suspect placed his left hand inside his cavernous shirt; he was reaching for something. He pulled out a large handgun.

We were both running furiously. My radio's microphone had come off my shirt collar and was now dangling between my legs, striking my knees like a pendulum as I ran.

It was surreal. I felt as if I were watching the scene from above instead of participating in it. Everything was moving in slow motion. Instead of seeing the dreaded orange flash followed by a loud explosion—and possibly getting shot—I watched his arm whip out, making the handgun go spinning through the air, landing in the parking lot behind the Salvadoran restaurant. He had tossed it.

Relief gave me a second wind. I was gaining on him, hoping to close the gap and tackle him. His hands were working at high speed as he ran. Like an expert multitasker, he was retrieving items from his clothing and tossing them into the bushes near the sidewalk; fortunately for me, that was slowing him down.

Suddenly my foot slipped in a small crevice of the sidewalk and my right knee buckled. I almost went down to the ground. I just managed to regain my balance as the suspect, now twenty yards ahead of me, turned left again, this time heading south on Vista Del Monte Avenue. My heart sank—he was getting away.

I couldn't afford to let him escape. This was my first chase on the streets of Los Angeles, the crucial race where the mind-worker scientist became a real street cop—or didn't. I couldn't let this dangerous criminal get away. On my radio, I could faintly hear Andy's distant voice radioing directions to backup units.

One cruiser was on its way, Code 3, to our location. I turned the corner. There was no sign of Andy; I did not know how far behind he was. I was on Vista Del Monte and Andy was on Valerio Street. The LAPD policy mandated not to lose sight of your partner in a foot chase; on the streets, this policy was often ignored. I was supposed to slow down until Andy caught up with me. But I didn't, because the suspect would get away, and that was not an acceptable option.

I needed to prove myself.

The suspect was across the street now, running on the sidewalk next to houses. Bushy citrus trees in the front yards swayed eerily in the wind. Dense oleanders lined the edges of the lawns. I saw his fleeting silhouette moving against the shrubbery. Suddenly he made a move, which could have helped him get away. He dropped to the ground and started to roll under the oleanders into a yard. He was planning to jump a few fences and disappear into the neighborhood. But the two seconds he lost by this maneuver were the two seconds I needed to catch up.

As soon as he began rolling under the bushes, I was on him. I heard myself shouting: "Don't move or I'll blow your head off."

These tough words surprised me. I had never made such a threat in my life before. Inside, I was hoping I wouldn't have to follow up on my threat. If push came to shove, I would probably have gotten in a wrestling match with him in the hopes that Andy would get here in time to help. The kid stopped rolling.

I was crouching over him with my handgun pointed at his head. He knew the drill: he had turned on his stomach, with his head pressed against the earth and facing away from me, his arms spread outward like a T—just as we always made dangerous felony suspects do before approaching them. His body was still, except for his rising and falling chest. A few tense seconds passed before Andy appeared behind me. He was panting. So was I.

Make that three of us.

Andy jumped on number three's back, grabbed his right wrist, twisted it, and brought it behind his back, making the suspect cry in pain as Andy retrieved his handcuffs.

"Run back and grab that gun!" Andy shouted. "Those two

assholes are still behind us." I was still gasping for air, but I remembered the gun tossed in the parking lot. I didn't want to leave until after I made sure that the suspect was handcuffed, but Andy wouldn't have it. "I have him under control!" Andy bellowed. "Run and get the gun. Watch out for the other two."

As I jogged back, retracing my steps, I reached down, grabbed the dangling microphone cord, pulled up the mike, and started broadcasting:

"Nine Adam twenty-three, we have one suspect in custody, four outstanding suspects, all male Hispanics, last seen on Van Nuys Boulevard at Valerio. We need perimeter set up for the outstanding suspects."

As I ran toward Valerio, a police car came flying up Van Nuys Boulevard, its emergency lights whirling and its siren blaring. It made a fast turn east on Valerio Street and sped away. A second cruiser raced past straight to the Van Nuys/Valerio street intersection, where it screeched to a halt. I could hear the distant sirens of several approaching police cars. Loud sirens, increasing in intensity, served as a reminder to the suspects to give up if they were engaged in a fight with cops.

Suddenly the police helicopter appeared overhead, bathing the area around me in light. A third police car came speeding up and stopped in front of me. Officers Collado and Sobel jumped out and ran toward me, shouting, "Are you okay? Where is your partner? Where is the suspect with the gun? Where are the rest? Which direction did they run?"

"We got one. Andy has him on Vista Del Monte. There were four suspects behind us on Van Nuys, I think they must have run east on Valerio. Let me see if the gun is still there."

I ran to the restaurant parking lot to look for the gun our suspect had tossed a few moments ago. I hoped that his associates had not found it. No cop likes gang members to be carrying guns. I walked around, scanning the ground, looking under the parked cars.

Finally I found it behind the left rear wheel of a brown Chevrolet sedan. It was a Tec-9 semiautomatic pistol with a large 25-bullet clip, a piece designed solely for mayhem, but then what else are guns made for? I picked it up carefully with a rubber glove I always carried in my pants pocket to handle sensitive evidence. Two more cruisers had pulled into the parking lot. I heard one unit after another announcing their arrival on the radio. The block was overrun by police cruisers.

The police helicopter pilot was moving officers to different blocks, setting up a perimeter to contain the two associates of the suspect we had captured. The lookouts would be hard to locate, as we didn't get a close look at them. The searchlight on the helicopter had lit up large areas of Valerio Street, swiftly sweeping in a 360-degree perimeter, illuminating the dark graffiti-marked alleys, exposing hidden corners and crannies, revealing the hiding spots behind garbage dumpsters and short walls.

As I was shouting over the din of the helicopter, giving details to the arriving officers, Andy emerged from Vista Del Monte Avenue with the handcuffed suspect. Andy, still out of breath and very angry, was screaming at him: "You fucking asshole!" Chasing an armed gang member can be agitating but one must never curse at the suspects.

Suspects always want to run and escape; there is nothing personal in this game. Once the handcuffs are on, all hostility should cease—no cursing, no insults, no force. Once the suspect is arrested, his welfare is the arresting cop's responsibility. Andy took criminal behavior as a personal affront. I took hold of the suspect, putting him in the nearest cruiser. I was going to ask one of the backup units to transport him to the Van Nuys jail while we looked for other suspects.

As I grabbed the suspect's arm, a flicker of recognition lit up my eyes. "Hey, Andy, isn't this our old friend? Sir, how nice meeting you again."

He had changed clothes, but this was definitely Alfredo—the robbery victim whose car had been trashed a few hours earlier in the 7-Eleven parking lot, literally steps away from the liquor store where my chase had begun.

Suddenly everything was crystal clear. Earlier today, Alfredo, a member of the San Fernando Gang, was caught by his rivals, the Valerio Street Gang, in their territory. Gang members don't like territorial violations. Alfredo got off easy but had still paid a price—his car and his ego were damaged. Tonight he and his four homeboys returned for revenge, carrying a 25-round semi-automatic pistol. When keen-eyed Andy spotted them, they had been scoping out the area, preparing to make a hit. We had prevented retaliation murders, albeit purely by accident.

This afternoon Andy and I had kept Alfredo from being killed by members of the Valerio Street Gang. Later in the evening, we had prevented Alfredo from killing Valerio Street gang members, along with some innocent persons. In a drive-by shooting, errant bullets often hit innocent bystanders.

Alfredo had had a tough day—almost killed, his car destroyed, and now off to jail. And what an exemplary day for Andy, whose internal crime radar and aggressive police work had prevented several possible homicides.

"Hey, Dutta!" Sergeant López was curious. "Did you run after him with your gun in your hand?"

"No," I answered.

"Do you think you would have had time to react if he shot at you?"

"I don't know," I replied softly and slowly, as if trying to justify my actions. "I just didn't think running with the gun in my hand was a good idea."

López gave me a quiet stare that got me thinking more than his words did.

———————

ANDY DROVE US BACK TO the station. I was completely drenched in sweat, my legs were shaking, my body was drained of energy—typical effects of the crash after an adrenaline rush. "Great job, partner! You saved some lives today!" Andy lightly punched my shoulder. "*You* are the one who sniffed out what they were going to do!" I gave him a weak smile. I was feeling melancholy, lost in thought, wondering about the significance of what we had gone through today.

The Valerio Street gang members were ready to kill to show their ownership of a street corner. Exhibiting control over their territory was a life-and-death issue for them, involving preservation of their honor and a display of their power. The Muslims, Sikhs, and Hindus behaved in a somewhat similar fashion but at a much grander, even genocidal scale, during the partition of India. Why can't we accommodate and assimilate? Why must we create lines and boundaries and differences, and then stop the "other" from stepping across these imaginary lines?

Alfredo had stepped across that line twice today. The first time he was the victim, the second, the persecutor. He was hell-bent to extract his revenge, even willing to indulge in mass murder to assuage his humiliation by the Valerio Streeters. Didn't Raju do exactly the same? First with Amar Singh, then with his French victims? Alfredo's life was saved; later some Valerio Streeters survived getting shot and killed because of this thin blue line separating the two groups—Andy and other police officers. Which thin blue line could have separated Raju from his victims or from himself?

We walked through the Van Nuys station; Alfredo was chained

to the pre-booking bench. Now that he was arrested, he looked happy, not a hint of worry on his face. For some reason, once they were arrested, gang members didn't seem burdened by the fear of jail anymore.

We took the elevator to the detectives' office on the second floor. Night detective Minton was working the desk. Detective Minton was probably in his sixties. He liked to dress in striped dress shirts and ties. With tortoiseshell reading glasses on the bridge of his nose at an angle, he looked like an inquisitive academic.

Criminals hated to cross Minton's path. He would dig through the penal code book like an IRS accountant and nail the suspects with the highest possible felony.

"So this is the same asshole who was the supposed victim earlier today! Since he returned armed the same day from Pacoima, to the same location where he was jumped, this gives us the intent. He was in a rival gang's neighborhood; he was armed with a loaded gun; he was dressed down like a gang member. I doubt he was there on a peace mission."

Minton narrowed his eyes behind his glasses and made his decision. "Book him for attempted murder and attach your robbery report to it in which he was the *victim* earlier today."

We booked Alfredo for attempted murder. Bail was set at half a million dollars.

It was two A.M. when I drove back home on empty streets and a wide-open freeway. Twice today I was moments away from random killings on Valerio Street. Had we missed Alfredo and his associates in the evening, they would have killed a few people, which in turn would have resulted in retaliation by the Valerio Street Gang. The punishment for straying in a different hood earlier during the day was murder, followed by more revenge and more murder. The cycle of revenge never ends; it keeps feeding on itself, getting larger, pulling in innocent victims, ravaging families, creating more monsters. And within that cycle, victims themselves turned into killers.

After the partition of India, both India and Pakistan became the aggressors in Kashmir and against each other, engaging in wars, slaughtering millions, endangering all of humanity with their nuclear brinksmanship—all this so that the Hindus could prove their superiority over the Muslims and so that the Muslims could prove they were better men than the Hindus; both sides standing ready to destroy themselves and the other to assert and defend their distorted sense of honor. Was there any difference between street gangs and military policy makers?

13

CRIMINALS? VICTIMS? DO WE HAVE A CHOICE?

*We must choose to live in this world and to project our own meaning
and value onto it in order to make sense of it. This means that people
are free and burdened by it, since with freedom there is a terrible, even
debilitating, responsibility to live and act authentically.*
—ALBERT CAMUS

My initiation in the LAPD changed me. I was exposed to a
new universe, exposed to a difficult, challenging, and ugly
face of humanity that few would ever see. All of my assumptions about people were tested; sometimes I felt like a misanthrope.
However, I began to settle into this new world, though I was forced
to develop two parallel lives: one in which I had a thickened skin as
insurance against daily pain and suffering; one in which I continued to feel in order to connect to and help the "regular" people, to
make sense of the human contradictions I had to confront daily.
My grounding in music and metaphysical poetry, my love for nature, my loving life partner, and my ability to step back and reflect
kept me from giving up and becoming cynical about humanity.

The selfishness and narcissism of the criminals I met were beyond any scale I could imagine. Every day I had to delve into a world of deviancy, making me wonder what caused people to be violent and sociopaths. Maybe if I could understand that, I could also understand what made Raju callous, predisposed toward revenge, and selfish; maybe it would help me deal with what our family had had to go through. Maybe I could wash away the shame and dishonor our family faced because of what Raju had done.

I have yet to meet a criminal who took full responsibility for his actions. Every criminal, including Raju, claims to be a victim: a victim of racism, society, poor parenting, abuse, a disadvantaged upbringing, a low socioeconomic status, mental illness, drug addiction, an unjust society. Do these factors cause deviancy? A cop would simply say that there are bad guys and there are good guys.

Was Raju predisposed to criminality or was he a victim of disadvantaged upbringing and poor parenting? Or are some people just born evil? His exposure to deviancy had begun early in his life. I still recall the day I stood in front of what looked like a sky-high large prison gate of the men's central jail in Jaipur. I was eleven years old and frightened; tears were welling in my eyes. We were visiting Raju. He was fifteen, arrested and imprisoned without a trial or formal charges, accused of being an enemy of the state!

Not many people remember the dark chapter in India's democratic history: the Indian Emergency of 1975. On June 26, 1975, the erratic autocrat, Prime Minister Indira Gandhi, had dissolved the Indian constitution and become the absolute monarch. The civil liberties granted by the constitution were revoked. The attorney general of India, Niren Dey, declared that under the emergency, "Every single policeman has the right to shoot and kill anyone on the road today. Nobody has the right to question." One of the world's most corrupt and unscrupulous police forces was given absolute powers to impose curfews and indefinitely detain citizens without cause. Tens of thousands of political opponents of

Mrs. Gandhi, including members of parliament and the leaders of opposition political parties, had been thrown in prisons. Informants were being paid money to turn in leaders of the opposition; the police were paying money for each arrest of those involved in demonstrations against Mrs. Gandhi. This is how Raju, apolitical and innocent of the charges, ended up in prison. Some unknown informant had earned money by implicating him as an enemy of the Indian state!

——————

THE GUARD PUT A BLACK stamp on my wrist before letting me in. I was trembling in fear as I stepped in behind my parents, realizing that I was inside a dreaded prison where only bad people from bad families end up. I knew that our family honor was in the dust, regardless of the truth that Raju was a victim, thrown wrongfully in a prison with hundreds of adults by a regime disregarding every precept of law.

We were ushered into a large courtyard. I saw him standing in the midst of a throng of white-clad politicians—a prison courtyard filled with hundreds of enemies of the state, everyone smiling and jovial! Raju appeared relaxed and nonchalant; he even bragged about how many famous political leaders he had met in the prison. Though innocent of the crime he was accused of, he had lost his innocence.

Nine years later, after he fled to France, I believed in Raju's ability to create a better life for himself and his family. He had a supportive wife, a job, responsibilities toward his wife and son. He was now a French citizen and did not have to return to India to face a long prison sentence. He was my sole sibling and I wanted to maintain a bond with him, even though it upset me greatly that he had never expressed any remorse for the suffering he had caused. Raju would periodically write or call from France, but I was turned off by his bitterness. He criticized everyone, believing he was better than others; he was not thankful for the great opportunity he had received to remake his life; and he always kept pressuring me to send him things, constantly asking for firearms! I began to shut him out of my life. I avoided returning his phone calls and rarely wrote him. Out of frustration and anger, I stepped away from his life. Then, one day, I received the call that he had become a murderer.

How does one graduate from stealing and conning people to killing in cold blood? Should I blame nature or nurture or just him—an individual unwilling to follow basic tenets of humanity? Or a combination of multiple complex facts? Nothing explains what Raju did.

———

ON JUNE 17, 2016, back home from the lung specialist's office, my life seemingly foreshortened, I had to take care of some urgent things. I called my commanding officer, Captain David Storaker. Dave is a tall genial garrulous man; we worked together as lieutenants at Topanga Division a few years ago. He knew me well; he was always helpful.

"I need your help! I don't know how much time I have left; I doubt much, but I don't qualify for a pension for another eighteen months! I need to find out how we can make sure Wes can get my pension."

I was not concerned about dying, I was worried about her. I didn't know how she would take my imminent death. How does one cope when someone you devoted decades of your life to starts dying in front of you, and all you can do is watch helplessly? The median survival time after a diagnosis of metastatic lung cancer is barely seven months; my lung cancer had spread to my bones. I knew I was not a data point in a study. But having been a biology researcher, I also understood science and statistics. Unless I had one of the genetic mutations for which a targeted treatment existed, I would be soon turning into a data point. Dave promised to see if he could ensure that Wes could get my pension after my passing.

Between the diagnosis and the lung biopsy results toward the end of June 2016, I got worse. I could barely walk, then was bedridden. I couldn't breathe. I was hooked up to an oxygen tank. I had never felt such excruciating pain; opioids were failing to even mask the pain. My dog Nanook lay down next to my bed and refused to move away, not even to relieve himself. Visitors had to step over his giant frame to get close to me.

My oncology team was planning to put me in a clinical trial to test a targeted chemotherapy medication in combination with an experimental immunotherapy drug. I needed one more test, a week, before the trial could begin. It did not happen.

On the evening of July 3, 2016, I lay semiconscious on my bed. I was certain I wouldn't survive the night. The pain was unbearable. I involuntarily moaned with every shallow breath I took.

"This may be my last night," I whispered to Wes. I was in an opioid-induced stupor. I didn't know if she could hear me; I couldn't see her reaction. I wrapped my arm around her, holding her tight. What more could I ask for? I was cycling in and out of consciousness. The opiates in my bloodstream were completely ineffective in masking the pain. The oxygen machine was making a cyclical buzzing sound, as the long tubes attached to it pushed oxygen in my nostrils. I could hear my moans. I was ashamed of this indignity, but helpless to stop it. Is this how death comes? I wondered. This labored breathing, this pain, and suddenly I will go to sleep forever and feel no pain? I would have gladly accepted sleep without pain; I had no objections to it.

Next morning, I was still around. Wes and her mom rushed me to the emergency room of the City of Hope hospital. It was the morning of July Fourth. As they wheeled me in, the hospital looked deserted. The nurse asked my pain level on a scale of ten. I smiled and feebly whispered, "Ten."

"At least you are smiling," he said cheerfully.

Three doctors were in the room. The primary one, Dr. Ravi Salgia, was an Indian American.

"How do you feel right now?" Salgia asked. I could barely breathe or speak, but I managed to recite a Persian couplet by the mystical poet Bedil Dehlavi:

"Tarq-e-arzoo kardam, ranj-e-hasti aasan shod
Sokht per feshani ha, qin qafas gulistan shod."

"You will have to translate it for me," he said. I had quoted what is one of my most favorite poems, something that reveals a little about my philosophy and happened to be perfectly apt for the moment:

"I cut off my ties with desire, suffering of existence became
    easier to bear;
I gave up the urge to spread my wings, my cage transformed
    into a garden!"

Dr. Salgia smiled; I passed out. Later, I found out that he had specifically asked to be my oncologist.

Years before I was diagnosed with cancer, I had come to see how vanity and anger had destroyed my family. False pride, vengeance, and illusions of victimhood kept my father's family forever mired in feuds and intrigue. Revenge continued to break them apart, until nothing was left. Avinash, my aunts, and my grandparents, who took out their angers and frustrations on Raju and me, splintered into their own opposing groups. My grandfather Bhapaji and his wife Bhabhiji turned against each other. Avinash left for his own home. Bhabhiji joined ranks with Avinash. Aunt Kanta and

Bhapaji made a front against others. Aunt Nanda turned against everyone else. The anger turned so bitter and pathological that after a loud shouting match with Nanda, Bhapaji stopped eating until he fell into a fatal coma. As Bhapaji died slowly, his wife did not even visit him. For the rest of their lives, Avinash and Nanda, my two tormentors, did not speak to each other, nor did they speak to Kanta. Kanta's disloyal daughter left her and moved in with Nanda. Kanta died alone, in misery, her dead body found by her neighbors the next day.

A photograph of us in a dictionary would be sufficient to exemplify a dysfunctional family. Nevertheless, I tried to bring them all together. In 2009, on a trip to India, I took my father to meet each one of his siblings in Jaipur. I was not expecting any reconciliation or about-face; I simply wanted them to let their anger go and stop being enemies. It was gratifying to see them talk, to recognize each other's humanity, after decades!

———

DESPITE THE FACT THAT CANCER had taken over both of my lungs and spread to my spine, a specific mutation in my cancer cells saved my life. A targeted chemotherapy existed to treat the mutation. I survived. The tumor shrank and there were fewer cancerous nodules. I eventually had enough energy to walk again, enjoy my time with family, visit with friends. Then in June 2017, the targeted therapy stopped working. The cancer cells had become wise to it and began to multiply rapidly; the cancer became aggressive and spread to my spine. Before Dr. Salgia shared the scan results with me, I already knew; I had been feeling short of breath and knew there could be only one reason for that. Usually cheerful and optimistic, Salgia was serious.

"Last July when you came here, had you been late by a day or two, it would have been too late," he said, gingerly touching on something we both knew but never mentioned. "You got one extra year of life."

I nodded my head. I knew exactly where he was going.

We discussed the options, which were limited. A brutal combination of chemo and radiation that might prolong my life by a few months but would destroy the quality of my life had already been scheduled.

"I want to be frank with you. I will not go through this treatment just for the sake of doing something. I don't want to spend the remaining few days of my life in misery just to live an extra three or four months." I still wanted to die with dignity—not attached to tubes, not inside a hospital, not enfeebled by a debilitating combination of chemotherapy and radiation.

"I will respect your decision," Dr. Salgia said. "Let's see what the biopsy shows us; let's see if we have any other options."

Once again, just as happened last year, I was bedridden and then hospitalized. Once again I accepted my fate without fear, complaints, or illusions, remaining in good spirits. Wes was with me all the time. She slept with me on the narrow hospital bed; we shared food from the same plate. I knew deep inside she was fearful, but she remained strong; how could she help me if she revealed her distress?

Once again a biopsy revealed a mutation in the cancer cells that could be targeted. A targeted chemotherapy for this mutation was approved only two years ago! Once again I responded. I am on my third life. There are no guarantees or expectations for a fourth. I have no interest in immortality or a mythical afterlife or a soul that lives forever. I have never taken death seriously; I am neither afraid of it nor awed by it; I will accept death without complaints or regrets, whenever it happens. But whatever time I have stolen from death, and whatever time I have remaining, I spend reflecting on my past. As Samuel Johnson once eloquently put it, "When a man knows he is to be hanged in a fortnight, it concentrates his mind wonderfully." Besides focusing my mind, I find my impending death most liberating! It has freed me up to confront my shortcomings honestly, to be more truthful to myself, to seek answers, and not to shy away from my uncomfortable shameful dishonorable past.

WHENEVER I ASKED RAJU TO explain why he engaged in crime, why he hurt others, he always blamed something or someone else. Poverty, abuse in childhood, our dysfunctional refugee family, a lack of support by our parents, betrayal by Amar Singh, insulting words from his French victims. I do not believe any of his explanations. I was brought up under the same circumstances; our micro- as well as macro-environments had been the same. I had been beaten and abused just like he was; I had lived through a harrowing life-shattering betrayal without hurting or killing anyone, without spending my remaining life plotting my revenge. Any reason offered by Raju to explain his behavior stands in stark contrast to how I lived my life! Abuse, deprivation, humiliation, betrayal—we had faced the exact same challenges. Yet we made diametrically opposed decisions in our lives. The fact that we did make different decisions, decisions that had radically different implications and outcomes, gives me hope that the world, despite all its hardships and schisms, *can* be a better place.

What could explain why our decisions and lives turned out to be so different? I believe a large factor is the cult of victimhood—claiming and convincing ourselves that we are victims and not resilient humans. Covering ourselves with a blanket of grievances, we use our victimhood as a justification for our own flaws or for our manipulation of others. If I am convinced that I have been wronged and humiliated, it is but a small step from that point to seek revenge and settle scores. By believing in the cult of revenge, by being selfish and self-centered, by giving in to anger and hatred instead of forgiveness and understanding, and by playing the victim, Raju continued down the slippery slope until he hit rock bottom. But events beyond our control *do not* define us. Indeed, life is cruel and unjust. For some, it is far more cruel and unjust. But that does not give us the license to hurt others. The oppression, shattering, splintering, failure, and hopelessness of my family did not require me to perpetuate any of it. But it did compel my brother!

I have not come to terms with Raju. I tried to understand why and how he could kill three defenseless humans. I repeatedly asked him to explain what had happened, why did he do what he did.

He did reply. His stories changed with time. His last explanation: His friends had made fun of him, insulted him because of his medical condition. He lost control and started shooting. Three people lay dead in the room when he was finished. To my father, he gave a slightly different version. His friend Marcel Rapallini had made fun of him because he had no money and no property; another woman always taunted him about his clothing, telling him he had no class. Having been psychologically upset as a result of his medical condition, he exploded in a fit of temper and gunned down three people, including Marcel Rapallini's wife.

I am going through one of the most physically and mentally challenging medical conditions anyone could go through. I have suffered worse humiliation, including a questioning of my integrity and courage; I was disgraced and punished by those who hated me in the LAPD for no reason except for my socially progressive beliefs. Forget lashing out, I did not even utter a curse word against those who humiliated me. So how can I trust his explanations?

His explanations have been blatant lies! He had executed a genial wheelchair-bound invalid, fifty-nine-year-old Marie-Claude Gaudet, with a single shot in the back of her head, *inside* her home. Fifteen days later, he sneaked inside the Rapallinis' home and killed them, stealing two weapons from their apartment. These homicides were cold-blooded, calculated executions. The night he killed the Rapallinis, minutes later he was nonchalantly having dinner with his wife and son. Raju's family had no idea what he had done.

I stopped trying to understand Raju's actions after corresponding with him for a short time. As I read his letters, it seemed as though I was reading a sociopath's manifesto. He had glib answers for everything. Everything he had done was someone else's fault; he just reacted to *his* victimization by *others*. Raju is serving a life term in a French prison for homicide. He recently learned from my mother that I have terminal lung cancer. He cried a lot. He offered to donate one of his lungs to me. This is the first time in my life Raju has ever offered anything to me. He wants to continue to talk and to maintain relations with me. I am not sure if I do.

The cult of victimhood and its progeny, self-righteousness, anger, and vengeance, made Raju what he is. Each day I worked as a police officer, I did my best to shape the attitudes of victims and criminals I met, to prevent them from falling prey to this cult. I still believe that forgiveness and compassion alone can make the world a better place and can help us escape the tentacles of anger and bitterness. My own life is living proof of that. But for whatever the reasons, I have not been able to forgive Raju.

Seemingly hopeless struggles which disadvantaged families face are *not* the reasons for people to become depraved or warped. Ignorance, prejudice, intolerance, terrorism, religious fanaticism, hatred, conflicts, wars, tides of refugees, and ever-shifting social dynamics of these times make it seem that we have no control over our circumstances. We need to know that we *do* have a choice in shaping our lives. Our choices matter. Every misfortune in our life provides us with an opportunity to improve the world around us, or, conversely, to destroy our own lives and those of our loved ones; we can bring paradise to our world, but we can also transform paradise into hell.

My brother and I faced exactly the same adversities but ended up creating two sharply divergent lives. What my life—especially when it is compared with my brother's—shows is that despite hopelessness, resentment, betrayals, and violence, love, dogged persistence, and forgiveness, admixed with humanity, help us overcome great odds and make life worth living. We do not choose our life's struggles, tragedies, and challenges; we do chose how we respond to them. Our choices about how we respond to adversity make us humans or sociopaths.

A life need not justify its existence, for as Ibrahim Zauq, an Urdu poet and the teacher of the last Indian emperor, said:

> Life brought us, we came; death takes us, we follow;
> The decision to arrive was not ours; nor the decision to
> leave.

But between the two involuntary points of birth and death, as Camus says, *responsibility to live and act authentically* is solely our own.

ACKNOWLEDGMENTS

This book would not have come into existence without the support and encouragement of Mary Bly. I want to offer my deep gratitude to Kim Witherspoon, my agent; William Callahan, for his help with the proposal; and Emma Janaskie, my editor at Ecco. In addition, people in many departments at Ecco, from art to marketing, have done brilliant work to get this book into readers' hands, and I'm very grateful to each of them.

Even given the help of these remarkable professionals, the book would never have been written without the support of those who have provided me with tangible, spiritual, and intellectual moorings, and created who I am—an extended family in India and America. I have been extraordinarily fortunate to gather loyal friends and mentors in the LAPD, in academia, and in the ranks of classical Indian musicians.

And finally, when I prepared myself for death two years ago, after receiving a diagnosis of Stage 4 lung and bone cancer, my brilliant oncologist, Dr. Salgia, promised me time . . . and I used that time to write this book. I am so deeply grateful for every moment.

CHAPTER 3 WAS PREVIOUSLY PUBLISHED in a modified form in Dutta, S. *Freedom, Partition, and Terrorism: Pakistan's Origin and Its Connection to Terrorism.* New Delhi: Har Ananad Publications, 2016.